Key Productivity and Performance Strategies to Advance Your Career

Key Productivity and Performance Strategies to Advance Your Career

Lesia L. Crumpton-Young, PhD
Vice President, Division of Research and Institutional Advancement,
Tennessee State University, Nashville, TN, United States

Academic Press is an imprint of Elsevier
125 London Wall, London EC2Y 5AS, United Kingdom
525 B Street, Suite 1650, San Diego, CA 92101, United States
50 Hampshire Street, 5th Floor, Cambridge, MA 02139, United States
The Boulevard, Langford Lane, Kidlington, Oxford OX5 1GB, United Kingdom

British Library Cataloguing-in-Publication Data
A catalogue record for this book is available from the British Library

Library of Congress Cataloging-in-Publication Data
A catalog record for this book is available from the Library of Congress

ISBN: 978-0-12-799956-2

For Information on all Academic Press publications
visit our website at https://www.elsevier.com/books-and-journals

Publisher: Mica Haley
Acquisition Editor: Mary Preap
Editorial Project Manager: Timothy Bennett
Production Project Manager: Punithavathy Govindaradjane
Cover Designer: Christian Bilbow

Typeset by MPS Limited, Chennai, India

CONTENTS

About the Author ... vii
Preface ... xi
Acknowledgments ... xiii
Introduction ... xv

Chapter 1 Strategies for Enhancing Productivity and Performance ... 1
1.1 Set Clear Goals, Objectives, and Targets ... 2
1.2 Prepare a Plan for Accomplishing Goals ... 9
1.3 Practice Strategic Immersion ... 17
1.4 Employ Decisive Decision-Making ... 21
1.5 Adopt a Get Things Done Mentality ... 23
1.6 Bind Thoughts of Fear, Doubt, and Worry ... 24
1.7 Overcome the Limiting Views in Your Mind ... 26
1.8 Silence the Critic ... 29
1.9 Chase the Fat Rabbits ... 33
1.10 Concentrate on Completing and Closing Matters ... 35

Chapter 2 Practices for Improving Productivity and Performance ... 39
2.1 Establish a Paradigm for Ensuring Consistency ... 39
2.2 Follow a Critical Path Analysis for Task Execution ... 42
2.3 Practice Efficient Allocation of Resources ... 44
2.4 Employ Technology Tools to Expedite Task Execution ... 48
2.5 Adhere to Practices That Promote Rest, Rejuvenation, and Stress Relief ... 49

Chapter 3 Conclusion ... 53

Productivity/Performance Log and Journal ... 55
References and Resources ... 75
Index ... 77

ABOUT THE AUTHOR

Dr. Young is the recipient of the United States Presidential Award for Excellence in Science, Mathematics, and Engineering Mentoring (PAESMEM), which she received from President Obama in 2010. She has worked with hundreds of individuals around the country on developing strategies and implementing practices to improve their productivity and performance. Currently, she serves as the Vice President of Research and Institutional Advancement as well as the Chief Research Officer at Tennessee State University. Dr. Crumpton-Young also serves as Director of the Center for Advancing Faculty Excellence (CAFÉ) that conducts research and provides professional development services for helping faculty and students improve their effectiveness, performance, and productivity. Dr. Young is a Certified Life and Career Coach who uses her knowledge and experience to help further the career of individuals throughout the nation. Dr. Crumpton-Young has coauthored a workbook entitled "Advancing Your Faculty Career" and authored the "You've Got The Power!" Workbook series dedicated to empowering individuals to unleash the greatness that exists within them. Dr. Young was the founder and former CEO of Powerful Education Technologies a company dedicated to enhancing the personal and professional development of youth and adults throughout our nation. Also, Dr. Young is the founder and served as Executive Director of the Power Promise Organization, a nonprofit entity dedicated to helping students realize the promise of a brighter future.

Previously, Dr. Crumpton-Young also served as the Associate Vice President at Tennessee State University and as a Program Director at the National Science Foundation. Also, she served as Associate Provost at Texas A&M University. She also served as Department Head and Professor of the Industrial Engineering and Management Systems Department at the University of Central Florida (UCF). At UCF she received the Trail Blazer award for being the first female to serve as a Department Head within the College of Engineering. Prior to joining UCF, she held the position of Associate Dean of

Engineering at Mississippi State University (MSU) where she was the first female to serve as Associate Dean of Engineering. Also, she was one of the first females to receive the Hearin-Hess Distinguished Professor of Engineering Award. Also, Dr. Crumpton-Young served as the Developer and Director of the Ergonomics/Human Factors Experimentation Laboratory during her tenure at Mississippi State University.

Dr. Crumpton-Young is a fellow in the African Scientific Institute. Dr. Crumpton-Young holds the distinction of being one of the first African-American females to reach the rank of Full Professor in Engineering in the country. She has served on the National Science Foundation (NSF) Committee on Equal Opportunities in Science and Engineering (CEOSE), the NSF Engineering Advisory Committee as well as the Army Science Board for our country. Dr. Crumpton-Young received her BS, MS, and PhD in Industrial Engineering from Texas A&M University; where she was the first African-American female to receive a PhD in engineering. Dr. Crumpton-Young received the 2017 STEM Innovators award as well as the 2006 Outstanding Women of Color in Science and Technology Educator Award. Also, she is the recipient of the 1999 Janice A. Lumpkin, Educator of the Year Golden Torch Award from the National Society of Black Engineers. Also, Dr. Lesia L. Crumpton-Young received the 1997 Black Engineer of the Year Education Award that is given to the one candidate whose qualifications place him/her in the ranks of the nation's highest achievers in the field of engineering.

In addition, Dr. Young is an active researcher and her various research interests include: Professional development, STEM education, mentoring, curriculum reform, STEM leadership development, human performance modeling and analysis, human reliability analysis, human fatigue assessment and modeling, use of virtual reality and computer simulation in ergonomics design and analysis, design of displays and controls, workplace design; carpal tunnel syndrome prevention and control; and workplace redesign for disabled persons. Dr. Crumpton-Young received the CAREER development award from the National Science Foundation for her research on Developing Engineering Criteria for the Inclusion of Persons with Disabilities. In addition, she received the outstanding industrial paper award for her research entitled *An Investigation of Cumulative Trauma Disorders in the*

Construction Industry from the International Occupational Ergonomics and Safety Conference. Dr. Young has served as Principal Investigator on numerous research projects and published hundreds of scholarly publications. Her research has been externally supported by the National Science Foundation, Office of Naval Research, NASA, and Department of Education; also, she has worked on many industrial research projects with sponsorship from companies such as UPS, IBM, Caterpillar, Intel, Garan Manufacturing, Southwest Airlines, and Lockheed Martin.

Dr. Young has been married for 27 years to Mr. Reginald Young and is the mother of two beautiful daughters, Mattlyn Young age 20, and Ashlee Young age 18.

PREFACE

The economic climate and professional expectations in our country have caused persons to have to do more job tasks than in previous years; thus, there is an increased desire among persons to find ways to get more task completed within their allotted work period. This book shares valuable knowledge and insight on practices used by high-performing individuals to enhance their productivity and performance. Use this book to discover strategies and practices for enhancing their productivity and performance. As a result of completing this book, readers will gain information on strategies for enhancing their efforts to get more tasks completed as well as practices for ensuring better performance. The strategies presented in this reference text are applicable to individuals from all professions. The strategies presented are those that when adopted can change one's mind-set and assist in the formation of thought patterns that will be useful in getting tasks completed successfully. Also, the strategies presented are meant to increase one's knowledge of how to be more productive. The practices shared in the book are intended to serve as action items or steps that can be employed to actively ensure that productivity and performance expectations are consistently met. In addition to a self-help tool, individual reading program, or independent reading assignment, this book can be used as a group resource, class text, supplemental text, or reference book within any personal or professional development program, course, or curriculum.

ACKNOWLEDGMENTS

Writing a book is truly a labor of love and I am grateful to many individuals who supported and assisted me along this journey. I am thankful for all of my career coaching clients whose experiences and stories served as the source of inspiration for this work. I am grateful to my family members, Reginald, Mattlyn, and Ashlee, who served as a constant reminder of why it was important for me to finish this book. I am thrilled to have "the world's greatest little sister," Delphine, who encouraged and motivated me throughout this process. I am tremendously indebted to Dr. Clara Young who served as my friend, motivator, editor, and accountability partner along this journey. Special thanks to Andrew who helped to successfully complete the final formatting. Thanks to Mary and the Elsevier team members for all of their assistance with this project. Special thanks to my lord God and savior Jesus Christ for ordering my steps along the journey of completing this book. I am especially grateful to all of the purchasers of this book, may your PRODUCTIVITY and PERFORMANCE be enhanced!

INTRODUCTION

Research has shown that many persons are achievement-oriented and task conscious; thus most persons have an innate desire to be highly productive in life. This desire to be a high-performing individual who is very productive stems from a variety of reasons. I learned more about these desires throughout my professional career, while serving as a life coach, career coach, or workshop facilitator to thousands of individuals. Results from one national survey I conducted revealed that persons typically desire to be more productive for the following reasons: (1) personal fulfillment; (2) monetary gain; (3) professional advancement; (4) societal impact; and (5) improved quality of life (i.e., less stress). These results demonstrate that productivity and performance are critical components of professional endeavors, advancement, and ultimately career success. Additionally, this data shows that one's level of productivity and performance also impacts their peace of mind and state of happiness. Given the importance and the impact of productivity and performance on one's personal and professional advancement, it is critical that steps be taken to increase one's knowledge of strategies and practices that can be evoked to enhance productivity and performance.

Productivity is a measure of how much one does, accomplishes, completes, or executes over a time period with a given amount of effort and energy expended. Productive individuals are those that complete the majority of their defined goals, objectives, or work tasks. For example, if an executive has five tasks to complete within a week and he/she gets 80% or more of his/her task completed successfully within that week, he/she is considered a productive person. If he/she gets more than 90% of their tasks completed then they are considered a highly productive person. If an individual completes less than 50% of their tasks, he/she may be considered less productive within the work environment. In working with my coaching clients, I have noticed that many individuals are unhappy and frustrated when they do not successfully complete 50%–60% of their expected tasks or activities. These individuals appear satisfied or content if they have accomplished approximately 75% of their tasks; while others appear happiest once they have completed as much as 85% or more of their expected goals or objectives within the specified time frame. In 2016 while conducting a national workshop, participants were asked to respond to a question regarding their opinion such as "rate how productive you are on a scale of 1–5 with one being the lowest and 5 being the highest." The results showed that only 6% of the persons felt that they were very productive and rated themselves a 5, most of the respondents rated themselves with a 3 or lower, which implied that they do not feel productive.

Performance is a measure of how well someone completes an action, such as a task, activity, process, or event. Performance is considered excellent when a person completes a task and it is done very well and exceeds all expectations. Performance is considered good when an individual completes a task in a timely manner and it is free of errors or mistakes. In general, most professionals have productivity targets and performance expectations they desire to meet. However, often many individuals find that regardless of how much time and energy they spend working; they are not consistently meeting their productivity goals or performance expectations. In the national surveys that I have conducted, individuals cite the following factors as some of the reasons for why they are not consistently meeting their productivity and performance goals: (1) interruptions; (2) procrastination; (3) overloaded; (4) stress; (5) poor time management; (6) poor guidance from others; (7) poor planning; (8) boredom; (9) lack of resources; (10) unorganized; (11) poor working relationship with others; (12) inconsistent; (13) failure to delegate; (14) not being strategic; and (15) unknowledgeable of how to enhance their productivity.

According to a survey that I conducted in 2013, 69% of people said that they waste time at work every day. Of this group, 34% said they waste 30 minutes or less each day; 24% said they waste between 30 and 60 minutes each day; and 11% said they waste several hours per day. Additionally, 80% or persons stated that they spend their time on websites, such as social media, online shopping, sports, travel, and entertainment, instead of working while at the office.

Frequently, persons who find themselves in this predicament are unsure what has caused their diminished state of productivity or performance and what strategies or tools can be employed to enhance this lack of productivity. The chapters in this book share valuable knowledge on strategies that can be adopted to form the basis of a strong foundation for understanding factors associated with increased productivity and quality performance. Also, critical information on practices such as action steps or implementation ideas is also presented in this book to ensure that the reader is aware of what habits need to be formed to positively impact their productivity and performance results. The strategies and practices described in the chapters are those frequently used by very productive high-performing individuals.

CHAPTER 1

Strategies for Enhancing Productivity and Performance

It is commonly alluded to that successful people think differently. It is also commonly said that "Whether you think you can, or you think you can't—you're right—Henry Ford." These two phrases illustrate the power of our thought processes and the influence that our thoughts have on our resulting actions. This concept is also true as it relates to our thoughts that dictate the resulting actions that shape our productivity and performance. Our thoughts and beliefs are extremely powerful in shaping the approaches taken and strategies used to ensure that we complete tasks in a quality manner. The next few chapters of this book are devoted to describing cognitive approaches that can be used to formulate strategies for successfully advancing one's productivity and performance. The collective adoption of these strategies can transform your thinking and influence the subsequent approaches utilized in productivity and performance. While I am aware that if you are hearing these strategies for the first time, you may be skeptical of their effectiveness; I encourage you to ignore your initial disbelief or inclination to disregard the strategy. Instead I implore you to simply try modifying your mindset to embrace the concepts being presented. As you begin to think about these concepts and rehearse their meaning in your mind, you will find greater receptivity to them and a deeper understanding of the meaning of each strategy. I strongly encourage you to take a week or two and think about each strategy presented, open your mind to the potential power of each concept, and begin developing plans to adopt these ideas. As you adopt the various ideas, I also encourage you to discuss them with others and employ "what-if analysis" to further analyze each idea. Lastly, I encourage you to begin testing the success of each strategy by implementing it within your plans of task planning and execution.

Key Productivity and Performance Strategies to Advance Your Career.
DOI: https://doi.org/10.1016/B978-0-12-799956-2.00001-2

1.1 SET CLEAR GOALS, OBJECTIVES, AND TARGETS

The road to success for ensuring productivity and performance begins with setting clear goals, objectives, and targets to pursue. A commonly made mistake is not establishing clear information on tasks to be pursued and activities to be undertaken; thus one may find theirselves working on tasks but not on the right tasks or doing things but not the right things that need to get done. It can be extremely frustrating and discouraging to find oneself working on tasks but not the appropriate ones that are well connected or aligned with the important goals. This action causes one to waste time, effort, energy, and other resources that should be conserved for use in completing tasks that are well aligned with the desired goals and outcomes.

It is often difficult to differentiate between goals, objectives, and targets because these words are sometimes used interchangeably to convey a similar concept, point, or idea. However, to ultimately ensure that one is productive, these three terms must be used to convey specific points and unique concepts. Goals can be defined as the overarching principles that guide the decisions or actions to ensure their attainment. Thus, as it relates to productivity and performance, a goal is defined as a primary desire or outcome that one wants to achieve. For example, getting a job promotion can be a goal. Completing a degree, course, or certificate program can also be defined as a goal. Also, buying a new car or a beach home can also be examples of goals. If one is in the sales profession, getting new customers or retaining existing customers can also be laudable goals. Taking the time to specify a goal is the most critical step to ensure its attainment. While this sounds like an elementary or fundamental principle, it is amazing how many people embark upon their daily activities without taking the time to specify a precise goal that they want to actively pursue with their time and energy; thus, they often find that a considerable amount of time has passed and energy has been expended but they have not achieved the outcomes that they truly desire. Thus it is critical to take the time and identify specific goals and desires that you have for yourself in all important areas of life. Examples of important areas or aspects of life include: personal development, career pursuits, financial aspirations, family/friends/relationship matters, spiritual growth, and health and wellness. Also, it is critical to identify both short-term and long-term goals for each important area of life. Because in modern society things appear to be extremely dynamic and change so rapidly, short-term goals are defined as those you would like to achieve over the next 1–12 months in the various areas of life where you place the greatest priority. In keeping with this time frame, long-term goal setting is defined as desires you would like to achieve over the next 13–36 months. While completing the goal-setting process it is important to list all goals but to select a manageable subset of goals to initially begin pursuing. This can be done by identifying which of this is your top priority (i.e., your primary interests) and which of this is lesser (i.e., your secondary interests). It is recommended that one selects no more than five short-term goals as a primary interest to pursue simultaneously within the next 12 months. During this time frame, and after the primary interests are achieved, simply identify other short-term goals, such as those that were initially secondary interests, to pursue as your primary interests. Because of the critical nature of completing short-term and long-term goal setting some examples have been provided in this chapter to allow you to begin the process. An example of initial short-term goal setting is shown in Fig. 1.1 and

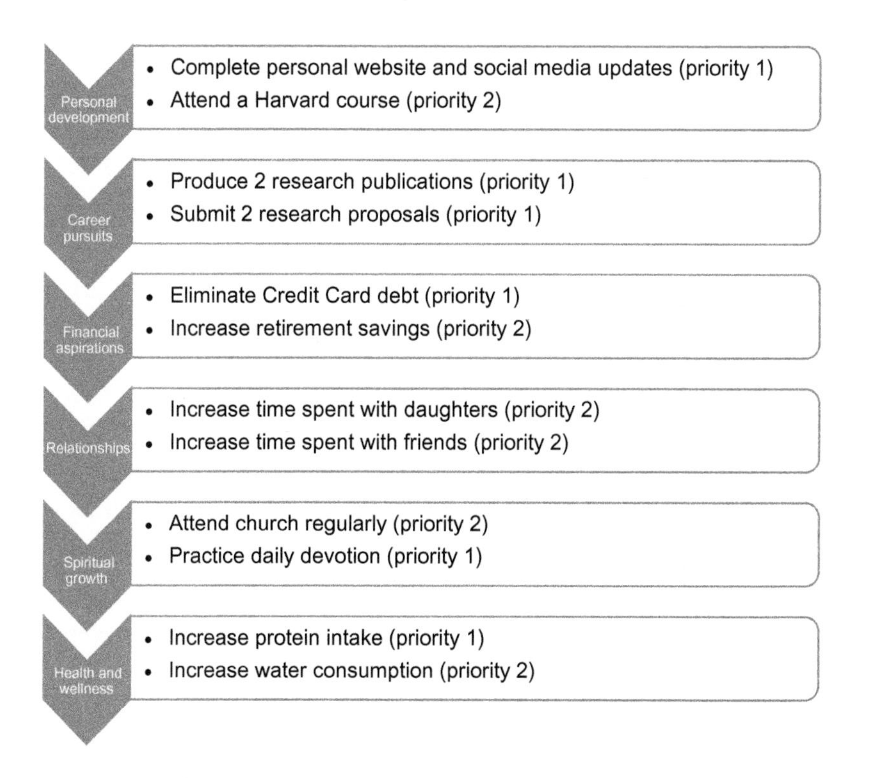

Figure 1.1 Example of short-term (1–12 months) goal setting.

long-term goal setting is shown in Fig. 1.2. Incomplete templates are shown in Figs. 1.3 and 1.4 to allow ample space for you to complete an initial goal-planning exercise. After you have completed your initial short-term and long-term goal-setting exercise, if you have more than five short-term goals listed, take the time to identify your primary and secondary interests. For example, of the short-term goals listed as an example in Fig. 1.1, note that each has been assigned a priority of 1 (i.e., primary interests) or 2 (i.e., secondary interests) as noted in the diagram.

Objectives are defined as the action or implementation steps that will be taken to achieve the short-term and long-term goals that have been identified as being important. To ensure success, it is advisable to define objectives in specific and measurable terms. Typically, objectives outline specifically "what or which" tasks need to be completed to

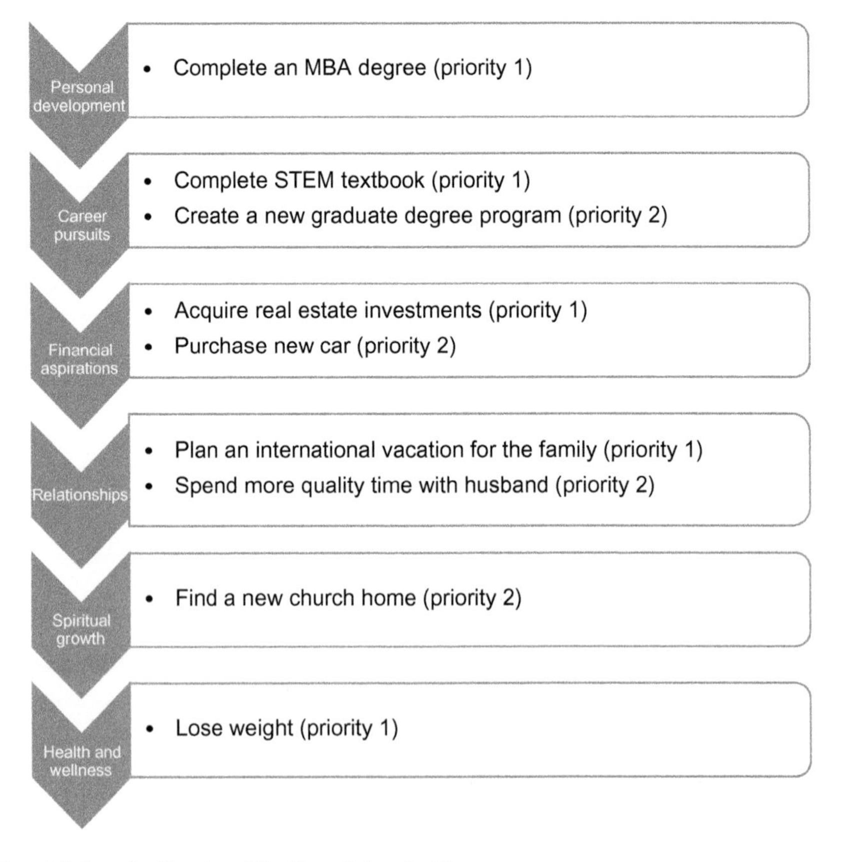

Figure 1.2 Example of long-term (13–36 months) goal setting.

Figure 1.3 Use for short-term (1–12 months) goal setting.

Figure 1.4 Use for long-term (13–36 months) goal setting.

reach a specific goal. If a goal is defined very broadly, objectives are defined or listed to serve as the blueprint in terms of how to achieve such a goal. The step of defining objectives to be pursued is a critical step for deconstructing goals into manageable components that can successfully be completed. Without identifying the key objectives, it is questionable if a goal will be achieved. Past experience with many coaching clients has shown that individuals are much less likely to achieve their goals if they forgo the planning process of clearly identifying the critical objectives that must be pursued. Also, the time to achieve a goal is more likely to be lengthened if proper planning of identifying the key objectives to be pursued is not done. Thus, to ensure that goals are achieved in a timely manner, clear objectives must be defined and executed fully in a timely fashion. Objectives are developed by simply listing all of the necessary action steps needed to successfully complete a goal. If there is uncertainty about which objectives need to be pursued to successfully achieve a goal, one can typically develop a list of needed action steps for most goals through conducting research, consulting experts, observing others, drawing upon similar experiences, strategizing, and thinking outside of the box. To illustrate the concept of developing objectives for the successful completion of short-term and long-term goals see Figs. 1.5 and 1.6. Fig. 1.5 exhibits the accompanying specific objectives which are action steps for each short-term goal listed in Fig. 1.1. Likewise Fig. 1.6 shows accompanying specific objectives which are action steps for each long-term goal listed in Fig. 1.2. After reviewing the examples in Figs. 1.5 and 1.6, use the sample templates shown in Figs. 1.7 and 1.8 to define the accompanying objectives for each short-term goal listed in Fig. 1.3 and each long-term goal listed in Fig. 1.4.

Targets are defined as the specific quantitative measurement, amount, or time frame for achieving a desired result. Setting targets and sticking to them are essential to ensuring consistent success in achieving productivity and performance goals. The short-term and long-term goals initially shown in Figs. 1.1 and 1.2 have been updated in Figs. 1.9 and 1.10 to include targets that serve as qualifiers to further define the desired outcomes of the goals. Also, the sample objectives shown in Figs. 1.5 and 1.6 for these short-term and long-term goals have been updated with specific targets as shown in Figs. 1.9 and 1.10. The establishment of targets helps to set expectations for achieving various goals and associated objectives; thus targets become a measure of quality that addresses a specified level of goal or objective attainment. Consequently, targets establish the framework for estimating performance which is a quantitative assessment of how well a goal was achieved, an objective was executed, or a task was completed. Thus through establishing targets for each goal and objective when

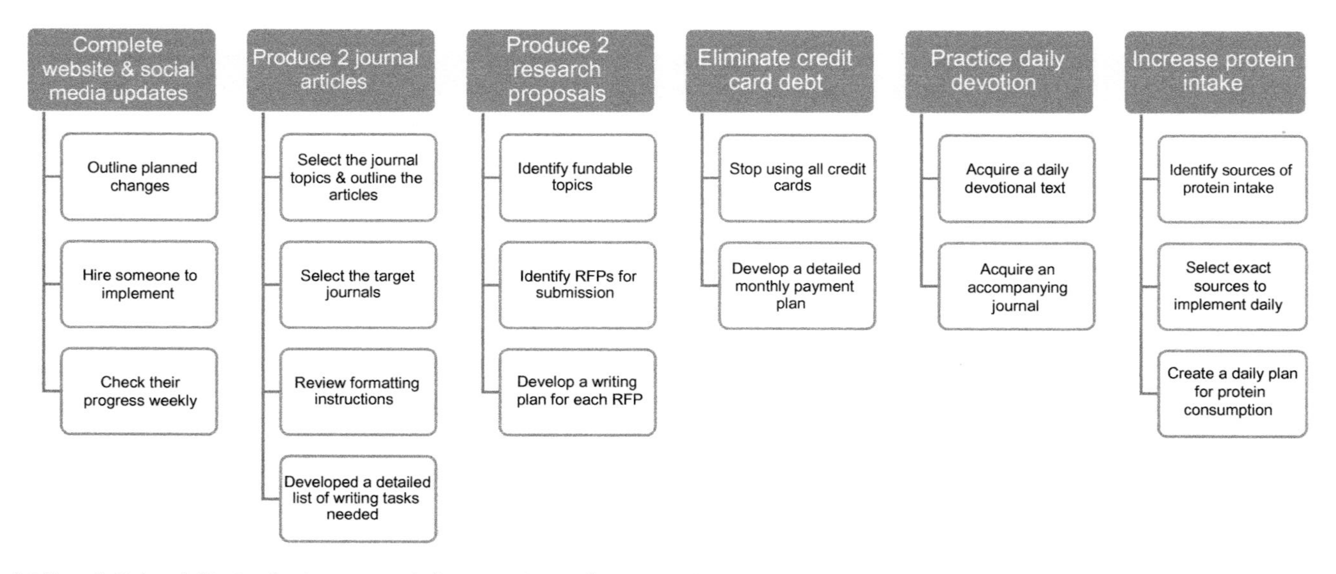

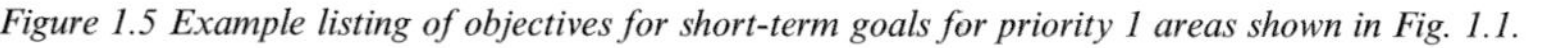

Figure 1.5 Example listing of objectives for short-term goals for priority 1 areas shown in Fig. 1.1.

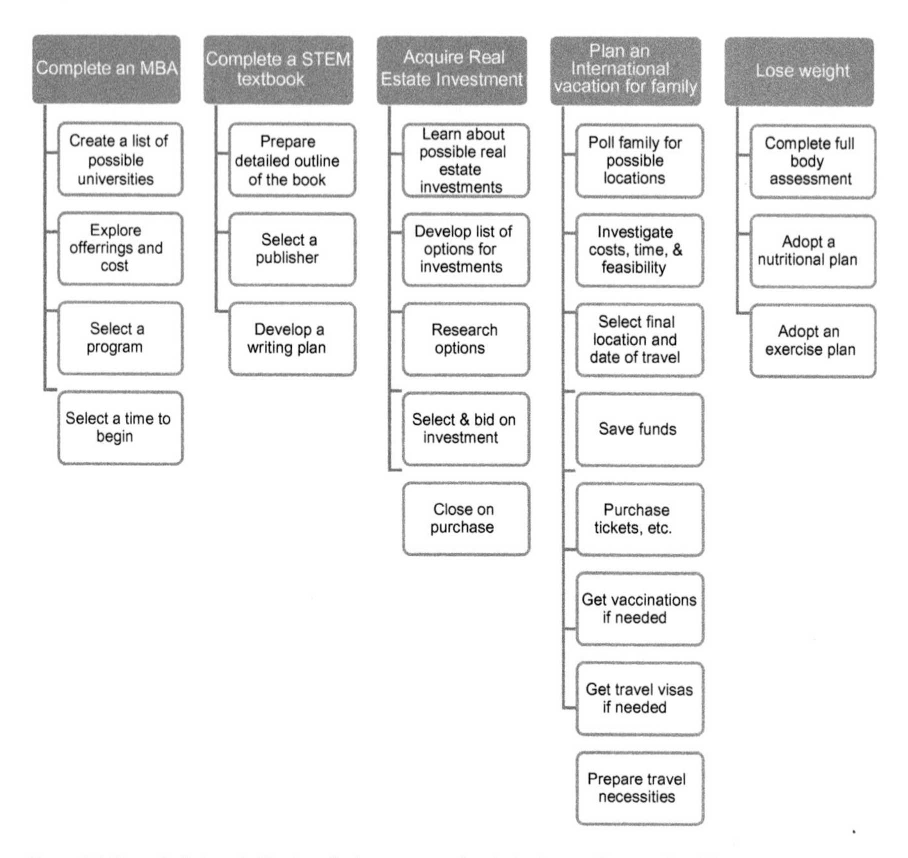

Figure 1.6 Example listing of objectives for long-term goals priority 1 area shown in Fig. 1.2.

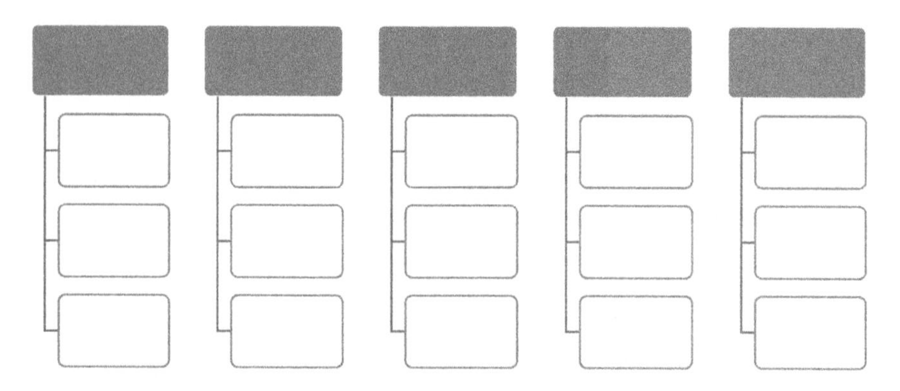

Figure 1.7 Template for developing a list of objectives to accompany YOUR short-term goals priority 1 areas shown in Fig. 1.3.

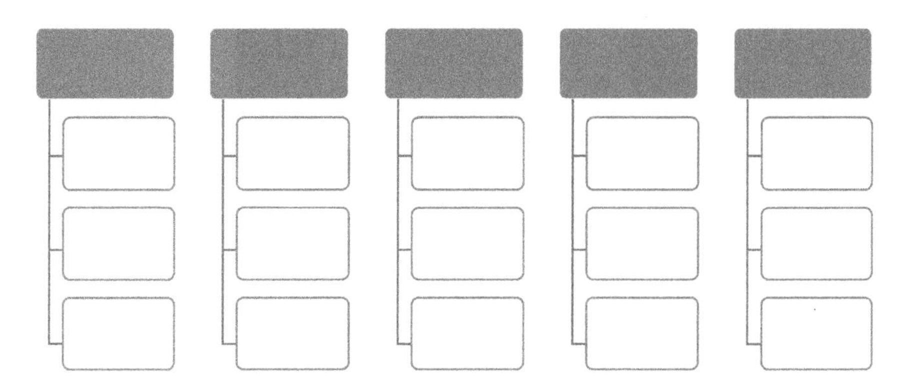

Figure 1.8 Template for developing a list of objectives to accompany YOUR long-term goals priority 1 areas shown in Fig. 1.4.

appropriate, one can accurately assess both productivity and performance.

Generally, it is gratifying and satisfying to achieve goals, objectives, and targets. Also it can be problematic if these aspirations are not achieved. For example, many people may feel down, depressed, and demotivated when targets are not met, objectives are not completed, or goals not achieved. As a result of not successfully accomplishing such wishes, one's self-esteem, self-value, self-worth, and self-image may suffer degradation. It follows from this degradation that one may experience periods of frustration, irritability, and negative thoughts. If this continues one frequently becomes dismayed, disengaged, and unhappy. If this is not handled properly it will continue to fester and one can begin taking self-disappointment out on others by being mean, angry, adversarial, critical, and judgmental. Exhibiting this behavior creates a situation where individuals are difficult to work with and toxic to be around; as a result, others may avoid interacting and spending time with such persons. To prevent the occurrence of these psychological and emotional issues as well as negative consequences, it is critical for individuals to set goals, objectives, and targets to help ensure that they are pleased with their productivity and performance.

1.2 PREPARE A PLAN FOR ACCOMPLISHING GOALS

After setting clear goals, objectives, and targets, the best strategy for ensuring attainment of these desires is to use a process for planning, scheduling, and managing task completion efforts. To effectively

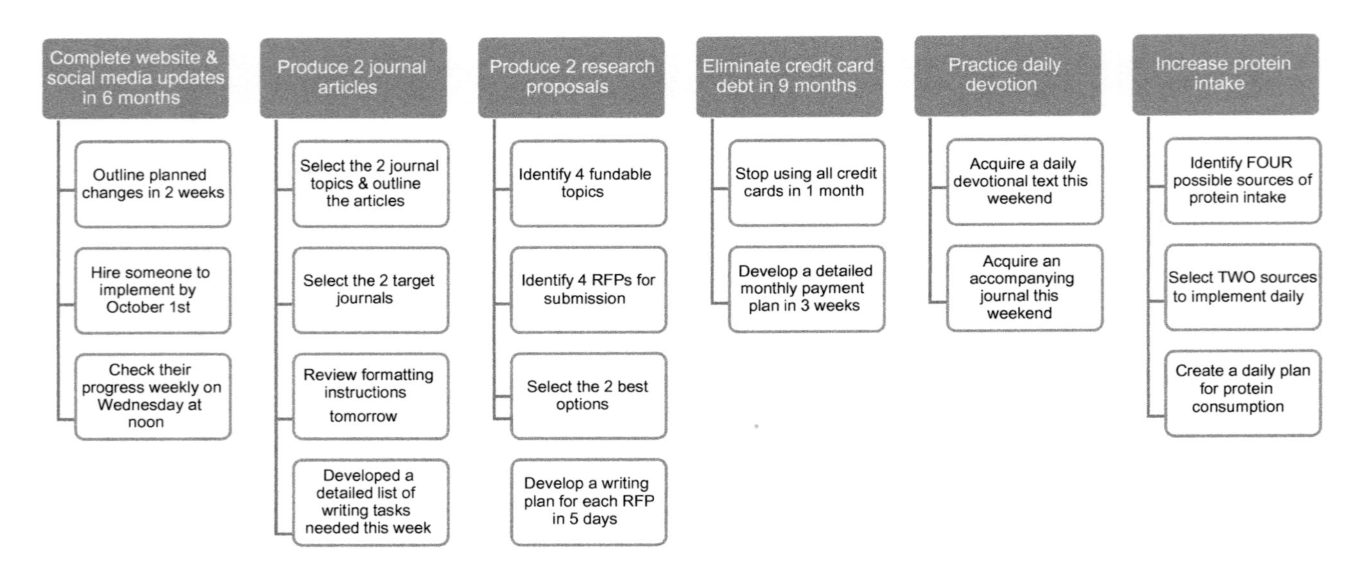

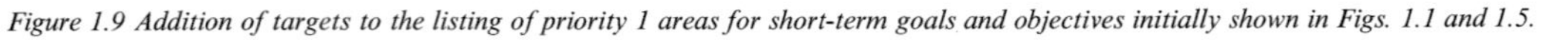

Figure 1.9 Addition of targets to the listing of priority 1 areas for short-term goals and objectives initially shown in Figs. 1.1 and 1.5.

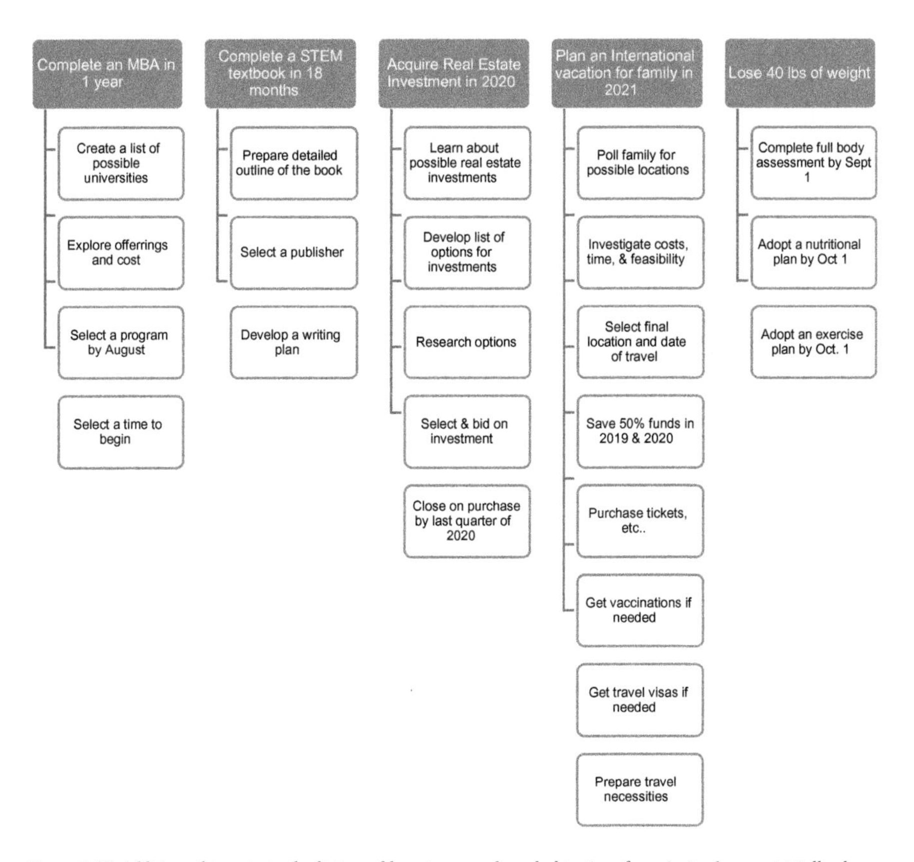

Figure 1.10 Addition of targets to the listing of long-term goals and objectives for priority 1 areas initially shown in Figs. 1.2 and 1.6.

ensure task completion, it is essential to use a planning and scheduling system. A well-designed planning system allows one to identify specific periods of the day when she or he will work on the various objectives or action items pertinent to the various goals is the desire to achieve. The goal-setting activity is the first crucial step to ensuring that one is productive; however, the planning of when one will devote time to complete various action items is the most critical component of ensuring that important task will be completed. Often one's level of productivity and performance suffers because they do not take the time to create a detailed plan of when they will work on certain action steps or tasks that must be executed to achieve attainment of multiple goals. After goal setting is completed, it is essential to immediately create a schedule when the necessary action items/tasks will be completed. Most planning systems fail because they do not allow individuals to immediately identify when during the course of their day, week, or month they will complete certain action items or task related to achieving desired goals. As a consequence, without this clear schedule, most persons will work on a variety of tasks throughout the day, week, or month but will not have a mechanism for ensuring that they are working on the most important tasks related to achieving their actual goals. Thus they are likely to realize that while they have been busy working, they have not worked on the critical tasks that matter the most with regard to their wants and desires. Hence without the identification of a plan detailing the allocation of time that determines when a task will be performed it is less likely to be completed. Inputs on behavior collected from coaching clients showed that those who developed a detailed plan of when they would work on completing tasks essential to important goals were 90% more productive than counterparts who worked for the same period of time. For example, those who did not develop a detailed plan were likely to complete 2–3 critical tasks per day. However, those coaching clients who developed detailed plans were more likely to complete 4–8 critical tasks per day. Coaching clients who utilized the detailed planning system were also less likely to feel "that they worked all day but didn't accomplish much" and more likely to feel content with their progress and reported that they had a "good day" when providing feedback. This research has shown that it is very frustrating and disheartening for persons when they have worked a full day but not accomplished very much of the important tasks that are related to their overall goals.

Data collected from various coaching clients reveal when asked, the majority (60%) of persons responded they do not develop a planning system because they do not have time to do so. However, experience has shown that actually by developing and using a planning system they are more productive in less time. Therefore, in reality, by using a planning system they are saving time and using it more efficiently, hence they are wasting less time. Thus in actuality those taking the time to develop a planning system are in fact saving time. Prior research and experience with highly productive persons has proven that "time spent planning is time well spent."

Also, research findings with various clients have shown that, when asked, many (30%) responded they are not sure of the best approach to use for planning and scheduling nor when they will complete the important tasks associated with their goals, because they are too busy doing all of the other tasks such as responding to emails, answering calls, meeting with colleagues, trouble-shooting problems, and managing conflicts. Persons who spend the majority of the day working in this mode typically find theirselves exhausted and disappointed with productivity and performance. Thus it is crucial to break this cycle by creating a plan of scheduling each day, week, or month to ensure that they have allocated time to work on the most important action items related to goals, objectives, and targets.

Creating such a schedule plan can be completed in three easy steps: (1) determine how much time per day, week, or month to allocate to each of the major goal areas of importance; (2) determine when to schedule the desired block of time identified in Step 1 into the daily, weekly, or monthly schedule; and (3) use the input from Steps 1 and 2 to draft a tentative plan of the schedule that will be followed to embark upon completing important tasks. An example follows to illustrate the ease of completing these three steps to generate a schedule plan.

Step 1: Determine how much time per day, week, or month will be allocated to each of the major goal areas of importance. To illustrate this point, each of the goal area listed in Fig. 1.1 is shown again in Fig. 1.11 with its associated estimate of how much time will be allocated to working on tasks associated with each objective of the major goal areas.

Step 2: Determine when to schedule the desired block of time identified in Step 1 into your daily, weekly, or monthly schedule. Past experience supports that when an individual identifies a precise block of time for completing a task it is more likely to be completed. Several persons have experienced this phenomenon and know it to be true. For example, if you need to meet with someone at work to handle a problem the meeting is more likely to occur if you actually schedule a certain date and time for the meeting to occur. This same principle holds true for completing objectives and tasks associated with accomplishing goals. Thus Step 2 of the planning process simply involves identifying blocks of time in your daily, weekly, or monthly schedule that will be dedicated to completing each objective or task related to each of the desired goals. In completing this step it is recommended to assign your priority 1 task objectives first. Priority 2 goals should be assigned specific blocks of time in the schedule only if time is available

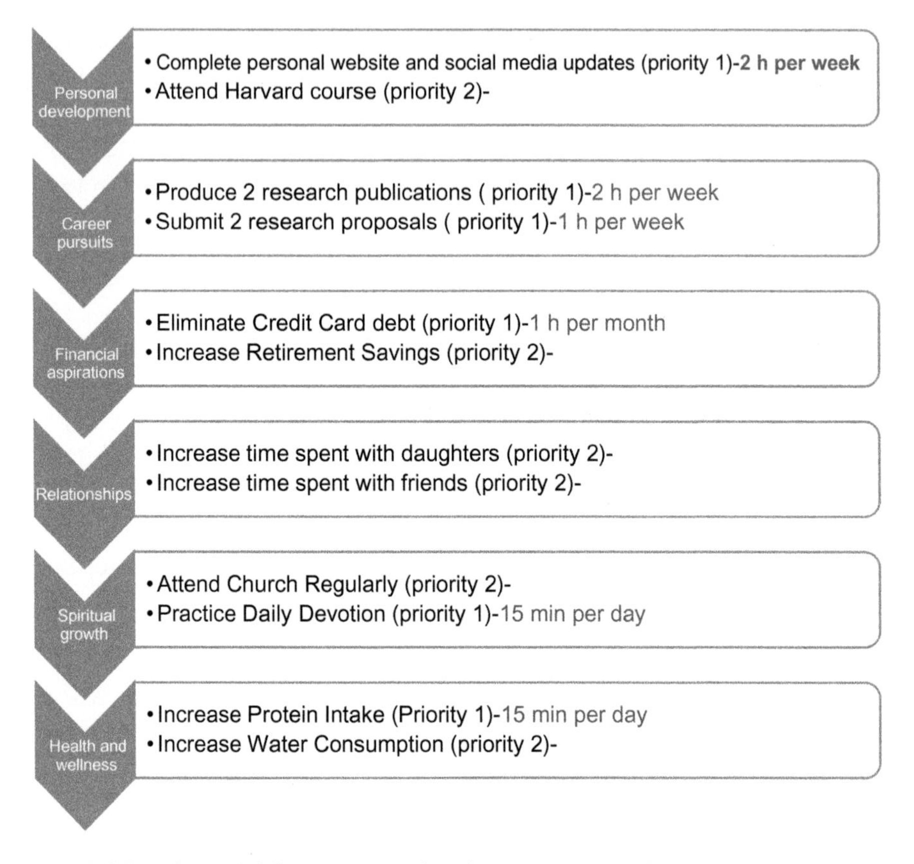

Figure 1.11 Example Step 1 of planning time to work on short-term priority 1 goals.

after assigning priority 1 objectives and tasks. When identifying days and times for scheduling these objectives and tasks, additional thought should be given to other factors that currently impact your schedule such as standing meetings, established routines, commitments, and preferences for working on certain types of tasks. To illustrate this point, each of the goals areas listed in Fig. 1.11 are shown again in Fig. 1.12 with a designated day and time that will be allocated to working on tasks associated with each objective of the major goal areas.

Step 3: Use the inputs from Steps 1 and 2 to draft a tentative plan of the schedule that will be followed to ensure completion of important tasks. When drafting this schedule take the time to include other factors that are an important part of your schedule such as traditional task activities, standing meetings, established routines, commitments, and preferences for working on certain types of tasks as shown in Fig. 1.13.

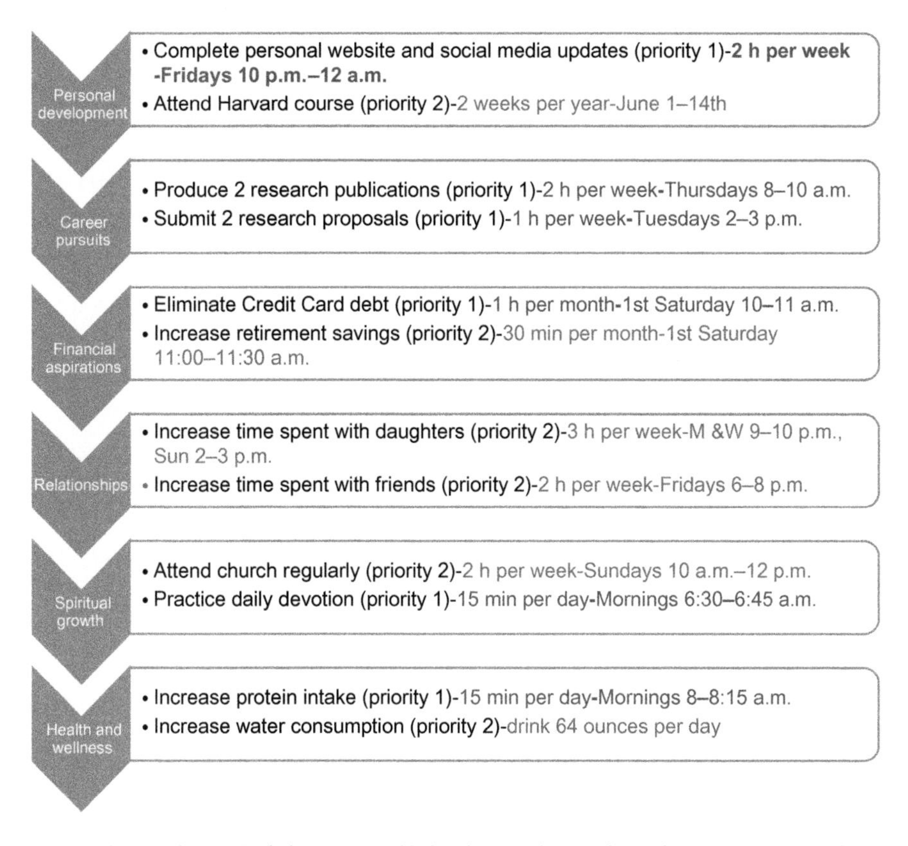

Figure 1.12 Example Step 2 of planning exact blocks of time to be used for working on short-term goals and objectives.

As illustrated in the earlier sections, the planning process for determining the amount of time and the specified block of time for executing objectives or tasks associated with major goals can be completed quite easily and efficiently. This planning process is an iterative process and can be changed or modified as needed to add time for completing important task objectives related to important goals you wish to purse. While this process is dynamic and can be changed easily it is important to ensure that the total time allocated for completing each task is consistent with the estimate of how much time should be allotted. For example, if your plan lists that you will spend 2 hours per week working on research publications then you must spend the 2 hours devoted to this task throughout the week, if you cannot work on the task on Thursday morning as planned in your tentative schedule it is perfectly acceptable to move it to another day and time during that week.

Time of Day	Mon	Tues	Wed	Thurs	Friday	Sat	Sun
6 a.m.	Daily devotion	Daily devotion	Daily devotion	Daily devotion	Daily devotion	Daily devotion	Daily devotion
7 a.m.	Morning routine	Morning routine	Morning routine	Morning routine	Morning routine	exercise	exercise
8 a.m.	email	Weekly cabinet meeting	email	Research publications	email	email	email
9 a.m.							
10 a.m.	Weekly directors meeting				Website & social media	Financial aspirations	Attend Church
11 a.m.							
12 p.m.							
1 p.m.							
2 p.m.		Research proposals					Activities with daughters
3 p.m.							
4 p.m.							
5 p.m.							
6 p.m.					Time with friends		
7 p.m.							
8 p.m.							
9 p.m.	Activities with daughters		Activities with daughters				
10 p.m.	Read & Rest	Read & Rest	Read & Rest	Read & Rest	Read & Rest	Read & Rest	

Figure 1.13 Example Step 3 weekly schedule of high priority tasks and activities.

However, remember that to ensure accomplishment of your priority objectives and tasks, the time allocation for each objective must be adhered to daily, weekly, and monthly as specified in the plan; thus if you cannot work on research proposals on Tuesday as planned this task must be performed on another day during the same week, avoid the inclination to simply forgo completing the task for 1 week in hopes of doing it the following week.

Preparing a plan for ensuring that time is allocated in your schedule to work on each of your priority goal areas is critical to your success in achieving the desired goals. Remember that "time spent planning is time well spent." As illustrated in this section, preparing a good plan can be done by following the three simple steps discussed previously. It is important to remember that once the plan is prepared, changes can be easily made to adjust the schedule as needed to accommodate the completion of other tasks. However, consistently following the plan is imperative to ensuring high productivity.

1.3 PRACTICE STRATEGIC IMMERSION

Strategic immersion refers to one's ability to be fully engaged in a task or activity while performing that task or activity. Being fully engaged helps to focus on the task and get it completed in a shorter period of time. Strategic immersion involves having both the mind and body focused on completing the tasks simultaneously. Mentally one has to block out other thoughts and keep the mind from wondering when completing the task. Thc busier a person is the more difficult it can be to mentally implement strategic immersion because there are so many thoughts of what needs to be done, what is undone, and the thoughts of juggling various demands. For example, have you ever been working on a planning schedule and yet making a phone call or answering a text message? Have you experienced a scenario where someone is meeting with you and answering the email at the same time? Have you seen persons who are typing on their computer, eating, and talking simultaneously? These are common examples of individuals who are not practicing strategic immersion and thus creating instances where the highest level of productivity and performance is not being achieved.

Often persons who are working in the scenarios above think that they are practicing high levels of productivity because they believe that "they are good at multitasking." However, they are mistaken because the scenarios above do not fit the traditional understanding of multitasking. Multitasking does not suggest doing multiple things simultaneously at the same time but simply being capable of managing multiple types of tasks. For example, individuals who are described as being good at multitasking may be those who can successfully complete many different types of tasks per hour, day, week, month, or year. Thus while the term multitasking refers to one's ability to complete various types of tasks over some total amount of time; the concept of strategic immersion describes one's ability to focus while completing each individual task. Experience shows that those coaching clients who practiced strategic immersion typically completed each of their tasks or activities quicker and with fewer errors than those who did not practice the art of strategic immersion.

Successfully implementing the concept of practicing strategic immersion is not easy for everyone and may be more difficult for some persons. However, it does become easier to achieve as one conditions their mind and body. The first key to success is to actively develop and use strategies for controlling your wandering mind. There are numerous examples of having a wandering mind. For example, have you ever found yourself reading a paper and yet the mind is thinking of other thoughts and as a result you keep having to reread a part of the paper? Or have you had the scenario where you are in a meeting or presentation but instead of listening intently to the speaker and thinking about the key points being made, your mind wanders to thoughts totally unrelated to the meeting or presentation being made at that time. Another example of a wandering mind may be that while you are working on your goal of writing a research publication, your mind may begin to think of the research proposal that you need to compete. There are numerous examples of how the mind wanders while one is working on various tasks, so do not become too alarmed or worry about this happening. As a result, it is not really a question of if your mind will wander, but more a question of what should be done when your mind wanders off of the current task being performed? Thus it is critical to develop strategies for becoming more aware of when your mind is wandering and how to successfully redirect the wandering mind.

Strategies for preventing a wandering mind include to begin each task with a statement that helps you to center your thoughts and begin to mentally focus on the task you are embarking upon. One strategy that I commonly use is to take a long deep breadth and recite the phrase "OK let's make this happen." Also, when working on tasks, some of my coaching clients have practiced strategic immersion by commonly adopting the mindset of telling themselves to "just focus for a short time so that you can accomplish the goal." Additionally, while striving to practice strategic immersion, some of my clients will commonly bargain with theirselves and say "just focus and stay on track with this task for 45 minutes and then you can take a 15 minute break." Remember it is not important which mental technique you use to practice strategic immersion, the most important fact is to actively develop and rehearse mental approaches to identify those that work effectively for you and allow you to focus and successfully complete your tasks.

It is important to ensure that strategic immersion is practiced with both your mind and body when working on a specific task necessary to achieve the highest priority goals, objectives, and targets. Just as there is a need to employ techniques for keeping your mind focused when working on tasks, there is also a need to keep your body engaged at all times while completing tasks. It is important to avoid the common trap of being idle during any time interval when you are engaged in completing a task. One also needs to minimize and avoid the urge to interrupt work by moving around, pacing, leaving the work area, and embarking upon other physical activities while completing a specific task. Therefore, the goal is to remain physically as well as mentally engaged. For example, if you are working on a task that requires typing, it is critical to keep both hands engaged to ensure the fastest completion time. The same idea is also true if you are cleaning beakers in the science laboratory, it is essential to keep both hands focused to get the maximum amount of the task finished in the amount of time spent working. Practicing strategic immersion helps with reducing the time it takes to compete a task by eliminating idle or unproductive time that typically extends the duration it takes to finish a task. By finishing a task in less time, it allows one to have more time available for other pursuits.

Also strategic immersion helps to ensure higher quality performance because fewer errors are committed while completing the task. Research and past experience has shown that common errors associated with tasks are the errors of omission and commission. Errors of omission are those that occur when a step is forgotten, and hence omitted, during task execution which typically happens when one is not as focused on implementing each of the steps required to complete a task. Commission errors are defined as doing something incorrectly while executing a specific task. The incidence rate of both types of errors increases if one is working without being strategically immersed in the task being performed. For example, if one is editing a paper while not being strategically immersed, it is very likely that a mistake will not be caught by the person who is editing.

While working toward increasing productivity and performance through practicing strategic immersion, there may be various factors that interfere with progress such as the belief that it is ok to become unfocused for a short period of time. Over the years, my coaching clients have shared that this is a dangerous belief, because their experiences have shown that after giving self-permission to be unfocused for a period of time, the actual time spent unfocused can be twice as much as expected. Hence, the decision to be unfocused for 30 minutes can easily become 60 minutes or longer and before realizing it you have been unfocused for a significant period of time which has been wasted. This decision to restrain from being unfocused is one of the most significant hurdles that must be overcome to ensure that strategic immersion regularly occurs. Also, the decision to not require oneself to get back to practicing strategic immersion after a period of time is another trait that can sabotage plans for advancing productivity and performance. Learning to recognize an unfocused period and forcing yourself to reengage in practicing strategic immersion is critical to long-term success of being more productive by completing more priority goals, objectives, and targets.

As discussed in this chapter, successfully practicing strategic immersion requires the use of active mental and physical strategies to control the mind and body. It is critical that you take the time to complete the exercise that follows where you develop a minimum of two strategies that will be used to ensure one's ability to actively remain strategically immersed, while performing various tasks or activities related to extended productivity and performance.

EXERCISE: Practice Strategic Immersion

Some practices that I plan to use to strengthen my strategic immersion skills include:

1.
2.
3.
4.

1.4 EMPLOY DECISIVE DECISION-MAKING

Decision-making can be difficult and often delayed because the difficulty of making decisions can cause a feeling of uncomfortableness. In addition, often various alternatives are pondered for too long instead of just pushing oneself to study the issues diligently and then arrive at a decision. This is the situation of analysis paralysis because instead of embarking upon the task the thoughts are simply buried; we are thinking and thinking instead of acting. The art of decisive decision-making is the practice of considering the most critical facts and information in a timely manner and expeditiously arriving at a decision. For example, if your goal is to produce two journal publications, one of the key decisions to be made is the selection of the journal where you would like to submit the manuscript for consideration. For several of my coaching clients they were significantly delayed in getting their articles published or never got it published because they couldn't make the decision of which journal periodical to patronize. As illustrated in this example, completing the task of selecting the journal periodical is a critical step in accomplishing the goal of getting journal article published. Thus it is vital to the overall level of productivity that we actively employ decisive decision-making techniques. Similarly, if one of your goals is to increase your protein intake, critical tasks or action items will include selecting which forms of protein you will add to your diet, what amount of each you will consume, and when through the day you will ingest the various protein options. I had one client who never achieved her overall goal of increasing her protein intake because she became stifled by the large number, variety, and complexity of the various protein options which made it very difficult to make a decision. Consequently, she became overwhelmed and discontinued her efforts to pursue her desired goal.

Also, employing decisive decision-making will help to reduce the long period of waiting and avoidance of making decisions as well as procrastinating on decisions; a trap that can commonly be fallen into while working toward achieving goals. As a result, it is imperative to handle matters as soon as possible and avoid delays associated with evaluating or assessing ideas and alternatives; thus allowing one to arrive at decisions quickly. Employing decisive decision-making techniques is applicable for most goal-related action steps especially those where a variety of options or alternatives are possible, such as our example of selecting which journal periodical to pursue or which protein option to execute. When the number of alternatives is vast making the decision difficult, the typical behavior of most persons is to simply delay making a decision and select the option of working on another unrelated task. By delaying that decision then it forces all steps that must be executed after the decision to be placed on hold as well; thus preventing the accomplishment of the overall goal, objective, or target. Similarly, for some individuals this dilemma may keep their mind occupied and inhibit them from moving on to other matters. When either of the two instances discussed earlier occur, it creates a situation that makes an individual highly inefficient. Thus it is critical that individuals employ decisive decision-making techniques to prevent confusion, frustration, stagnation, and other occurrences that will disrupt productivity and performance.

While working with various coaching clients, I created a decisive decision-making technique that worked well when multiple selection factors need to be considered to evaluate multiple alternatives. This technique was developed utilizing some of the basic principles of decision theory. Also, this is a simple, easy to use, time-saving method for making informed decisions. The essence of the technique is to identify the top five most important analysis criteria that will be used to evaluate the best five alternative options pertinent to the decision being made. This information is then placed into a decision table that will allow comparisons and contrasts to be easily considered. For example, when completing the action item of deciding which journal periodical to patronize when wanting to produce a journal publication, one could develop a decision table as shown in Table 1.1.

In this example, if all of the analysis criteria factors were equally important then alternative 4 would be selected; also if utilizing the least amount of costs were the most important criteria then alternative 4

Table 1.1 Example Decision Table for Analyzing Multiple Criteria and Alternatives

Top Five Analysis Criteria	Alternative 1	Alternative 2	Alternative 3	Alternative 4	Alternative 5
Costs	1000	400	600	300	500
Review time duration (months)	6	7	7	7	7
Manuscript length (pages)	25	25	25	25	25
Time before final print (months)	12	14	13	14	14

would be selected because it requires the least amount of money. However, if total time were the most important factor then alternative 1 would be selected because the manuscript would be published in the least amount of time. As demonstrated earlier this decision table is a simple technique that can be used to ensure decisive decision-making when completing intermediate tasks related to accomplishing goals, objectives, and targets; thus decreasing the time associated with making decisions and increasing one's overall productivity.

1.5 ADOPT A GET THINGS DONE MENTALITY

Having the cognitive frame of mind necessary to push and propel one to simply embark upon completing necessary tasks that we need to get done is essential to being productive. Often we find ourselves thinking about reasons why we cannot embark upon completing a task. Perhaps we do not feel up to doing it at the time. Perhaps we are distracted by other thoughts or individuals. Maybe we are waiting until we feel like doing it. Wc may also be waiting for someone else to take the lead or for the perfect moment or time to embark upon the task. Unfortunately, the perfect person, time, or situation may never present itself; and we may find ourselves waiting perpetually and become somewhat stuck in place. Productivity only occurs when we act and get the tasks completed. Thus it is critical to adopt the mentality of just getting it done, getting started, and making things happen. To be successful at being productive one simply has to alter their mindset to that of the one who is swift to act when it comes to getting started or finishing tasks.

Adopting a completion mindset means refusing to allow the common hindrances and distractions to occupy your mind and prevent you from embarking upon completing tasks. Successfully completing a task is reliant upon simply beginning the task and having the mindset that encourages you to get started working as quickly as possible. Adopting a mindset to just get started with a task is harder than it sounds for some individuals. Changing your mindset to one of action is what will ensure long term that you are someone who will get things done as opposed to someone who will simply think about getting things done, or someone who will wish that they could get things done. Our mind is very powerful, as an old adage says "whether you think you can or can't, it will be so." This saying illuminates the power of the mind and the importance of adopting the get it done mentality. The key to changing your mindset is to begin to actively think about and practice the art of conjuring up action-oriented thoughts such as "once I start this task, after I complete this task, just go ahead and get started, don't keep waiting, why aren't you making this happen." It takes an enormous amount of practice to develop this type of mindset.

Additionally, consistency is very important when working to change your mindset. It takes daily effort and activity to make permanent changes. As a successful farmer once said, the key to successful farming is to till some soil every day or move some dirt each day. If we are inconsistent then we would not realize the permanent change we are seeking. We will simply become frustrated by the lack of progress and resort back to our old habits which would not change our mindset. Consistency and commitment strategies are necessary to create the habits that form the new mindset of "getting task completed which is the foundational element of being productive." Consistency often requires a mental strategy that simply allows one to get started on a task so that progression can be made toward completion. It is like starting a fire, one small spark can get things going. Similarly, initiating the effort is the most difficult part. Think of a sail boat, one simply needs to initially get the sail up on the boat to catch the wind which continues to propel the boat; thus completing the task of sailing. Actively rehearsing thoughts or creating mental images of executing tasks will help to begin changing your mindset. Adopting the get it done mentality is essential to increasing your level of productivity.

1.6 BIND THOUGHTS OF FEAR, DOUBT, AND WORRY

As you embark upon executing tasks related to accomplishing goals, objectives, and targets, thoughts of fear, doubt, and worry will occasionally occupy your mind. Fear usually tries to convince us that something bad, negative, or different from what we expect will occur in the pursuit of being productive. Thoughts regarding poor performance or negative outcomes can hinder the pursuit of dreams or aspirations in life. For example, perhaps you are very interested in writing a book. Thoughts of fear will try to convince you that no one may buy the book or if you want to get a new job, fear may try to convince you that no one will ever hire you so don't bother with trying.

Slightly different from fear, thoughts of doubt will try to convince you that you are not capable or knowledgeable enough to accomplish a task or idea that you may have. Once again, if you are thinking of writing a book, thoughts of doubt may send negative messages to you, such as thoughts that you are not really a good writer because you do not have excellent writing skills or because you have never written a book. Or while trying to get a new job, thoughts of doubt may cause you to believe you would not do well in the interview.

Likewise, thoughts of worry will try to convince you that something will go wrong, or things will not progress as planned, and that problems will occur. Once again let us assume that you have an idea to write a book. Worrying will produce thoughts such as "You'll never finish this book" or "You'll never find a publisher." Or in trying to get a new job, thoughts of worry may make you think that you may get the job but would not be able to keep the job.

To ensure productivity and performance it is critical to silence concerns of fear, doubt, and worry. These three phenomena can hinder your progress of embarking upon and completing tasks. Paralysis happens because one becomes afraid of the task that needs to be done, and/or the outcome of the tasks, and hence overthinks or contemplates how to proceed for too long and thus delays the start of the task. While coaching many clients over the years, it was discovered that the most effective mental strategies for responding to fear, doubt, or worry include: (1) ignoring the fear, doubt, or worry; (2) overcoming the fear, doubt, or worry; and (3) persisting in spite of fear, doubt, and worry.

Strategies for ignoring fear, doubt, and worry involve shifting your frame of mind to another topic when you encounter thoughts of fear, doubt, or worry. For example, if you are working on writing a journal article and encounter thoughts of fear, doubt, or worry you can simply shift your mind to think about the beneficial aspect of the research and why it is important to share those results with the reader through publishing the findings. Also, if you are actively working on the goal of increasing your protein intake and thoughts of fear, doubt, or worry began to distract you from moving forward to implement changes then you may want to simply begin to envision how the desired changes will appear once you have met the goal of increasing your protein intake.

Likewise, strategies for overcoming fear, doubt, and worry begin with confronting the intimidation caused by these three phenomena. It is commonly stated that "FEAR = False Evidence Appearing Real." Thus an effective strategy to overcoming fear is to challenge the false information with factual data related to the topic. For example, if you begin to doubt whether you will ever complete the task of writing a journal article, this thought can be challenged by reviewing data facts showing that many other individuals have successfully completed this task.

Strategies for persistence in the face of fear, doubt, and worry require the creation of thoughts that allow you to move forward in the presence of fear, doubt, and worry in spite of your concerns. This means recognizing that these phenomena are present but having the resolve to persist in spite of their presence. Having this resolve simply means taking the first step to begin writing the journal article one page at a time in spite of the fear, doubt, or worry. Invoking various strategies for ignoring, overcoming, and persisting in spite of fear, doubt, or worry is vital to your productivity and performance.

EXERCISE: Binding fear, doubt, and worry

As you think of the goals, objectives, and targets that you are interested in attaining, list the strategies you will employ to:

1. *IGNORE fear, doubt, or worry:*
2. *OVERCOME fear, doubt, or worry*
3. *PERSIST through fear, doubt, or worry*

1.7 OVERCOME THE LIMITING VIEWS IN YOUR MIND

Often there are thoughts and views that limit efforts to get tasks completed which undermine efforts to be highly productive. For example, there may be thoughts in your mind that suggest that you cannot accomplish a goal because it is too lofty. Also, you may encounter thoughts suggesting that because none of your friends or colleagues have achieved their identified goals, perhaps you would not either. As a result of limiting views in your mind, if you convince yourself that a goal would not successfully be completed, then there is a tendency to develop a perspective of "why bother trying" which derails any efforts to achieve your goals. Also limiting views of capabilities may cause one to settle for less than their best performance. For example, this commonly happens when taking a course that is challenging, instead of working confidently to earn a top grade one may settle for earning a satisfactory grade. A limiting view of one's capabilities will also undermine one's competitive spirit which lowers their expectations of winning causing them to settle for something less before the competition even begins and thus diminishes their efforts to win the competition. In the aforementioned examples it is as if sabotaging one's success. Under these conditions of having limiting views it causes one to be their own worst enemy.

The concept of having limiting mental views means to have restraining thoughts that control actions and typically diminish achievements. There are several types of restraining thoughts such as opinions and beliefs. For example, if you believe that odds are stacked against you then you will exhibit behavior that does not exert our highest energy level. Also, limiting assessments of others or situations can also manifest as restraining thoughts. For example, if the belief is "women will always earn less money because they have in the past" then settling for earning less money without negotiating for an equal salary may be the outcome. Also, a limiting view may taint our analyses of whether we can accomplish the goal of producing two journal publications and convince use to constrain our vision to include producing only one journal publication.

Often the limiting views originate from various sources such as negative thoughts about one's self-worth or self-value which will ultimately affect one's self-image and self-esteem. Additionally, negative thoughts about one's preparation, experience, and/or expertise can undermine the successful completion of goals, objectives, and targets. Likewise limiting views about the knowledge possessed, skills acquired, ability to get task completed, and information on past performance will restrict future productivity and performance pursuits.

Recognizing the occurrence of limiting views may be difficult, especially if you have been experiencing them for a long period of time. However, research with many coaching clients over the years reveals that there are distinguishing characteristics of these limiting views that can be detected. For example, if the view in your mind creates thoughts that make you feel less knowledgeable, prepared, capable, or successful this can be classified as a controlling thought that you should not listen to at all. Also, if the thought in your mind restrains you from feeling as if you can accomplish certain pursuits it may be another form of a limiting view. In addition, if the thought in your mind regulates or reduces the excitement, energy, or enthusiasm that you have for embarking upon an endeavor then it may be considered a limiting view as well.

Additionally, if the thought is one that restricts or constrains the amount of effort that you are willing to put into completing the task or the expectations that you hope to gain then it is also a limiting view. While it may be difficult, it is critical to identify the limiting views in your mind to minimize the impact on your level of productivity and performance.

It is essential that strategies are developed for actively overcoming the limiting views in your mind as they occur. Several coaching clients who have experienced limiting views in their minds have used the following techniques to combat these limiting views:

1. Develop an exhaustive list of strengths and recite this list in their minds each time they encounter a limiting thought;
2. Describe some instances of past successes and rereads these examples when confronted with restraining thoughts;
3. Immediately create a positive thought/story in their mind to counter the limiting thoughts; and
4. Review pictures or examples of the success of others' that dispel the negative thoughts or limiting views.

Also, many coaching clients discovered that one can significantly minimize the occurrence of limiting views in their minds by regularly using the following techniques:

1. Beginning the day with a short motivational message;
2. Memorizing a positive quote for the day;
3. Regularly reflecting upon past accomplishments; and
4. Ending each day with noting the successes achieved.

In summary, the occurrence of limiting views in your mind can significantly affect your ability to be productive. To combat the impact of these negative thoughts, first learn to recognize them and implement immediate strategies to combat the messages that are being propagated in your mind. Next, we must develop strategies for reversing these sabotaging thoughts and implement techniques that will reduce and ultimately eliminate their occurrence. It is only through actively minimizing and eliminating the occurrence of limiting views in your mind one can maximize their productivity and performance.

EXERCISE: Overcoming the Limiting Views in Your Mind

Please complete the following statements:

1. *The limiting views that I typically experience include:*
2. *The strategy that I plan to invoke the next time I experience one of these limiting views include:*
3. *The technique I plan to implement for permanently reducing the occurrence of limiting views is as follows:*

1.8 SILENCE THE CRITIC

In today's society people can be extremely critical and this unfavorable nature can serve as a distraction and deterrence to productivity. Having a critical perspective can hinder thoughts and actions which create a paralysis and prevent embarking upon tasks. Also having an overly judgmental perspective in life creates enough criticism such that one can begin to undermine decisions and plans which then delays the progress by diverting attention and subsequent actions away from pursuing desired goals, objectives, and targets. Thus an internal critic can convince you that you are not going to achieve any of the goals set so why bother with setting goals at all. Likewise, this internal critic can convince you that planning is a waste of time because you will not develop a good plan nor follow it once it is created.

By definition, the critic is the inner voice that forms a pattern of destructive thoughts toward various situations, circumstances, and others. For example, the critic within may suggest that you are not good enough in Math so why try pursuing a career in Engineering. This inner voice can be extremely judgmental and cast a negative perspective on most scenarios; thus convincing you to arrive at thoughts or conclusions that serve to sabotage your efforts to be productive. Also, the critic within may pass judgment on your situation without thoughts being supported by facts and rational details. For example, if you set a goal to write a research proposal to secure funding, the inner critic may convince you that funding will never be received, because the agencies are biased in many ways; thus thwarting efforts to embark upon the task of writing a proposal. Also, the critic may create untrue thoughts that divert one's attention away from the most important goals, objectives, and targets that need to be pursued. This is particularly concerning because as one's attention is diverted to tasks not closely connected to goals, then you may find yourself busy working on tasks that do not move you closer to achieving the desired goal, objectives, or targets. I have experienced this scenario often with various coaching clients who were committed to working on producing two journal publications, but the inner voice convinced them that it is impossible to accomplish this task so they began to believe the voice and inadvertently turned their time and attention to working on other tasks and later realized they had not attempted to work on the journal articles at all.

The inner critic is powerful because it occupies part of your mindset and thus is very convincing. The inner critic knows your thoughts, values, and beliefs very well. The inner critic also knows your fears, doubts, and worries. Likewise, the inner critic also knows your insecurities. Because the inner critic has been a part of you for so long it also knows the mistakes that have been made as well as the disappointments experienced. The inner critic knows your hopes and dreams as well as strengths and weaknesses. In addition, the inner critic knows your likes and dislikes as well as comfort zone and boundaries that have been set for yourself in life.

As described in this section, the inner critic is very powerful because it knows so much about us and uses this information to manipulate our thoughts and actions. Thus it is important to identify the inner critic and develop some defensive mechanisms to diffuse its effect on your behavior and resulting actions. The inner critic can be identified by becoming aware of thoughts that you may have that cast negative judgments around your beliefs, thoughts, desires, endeavors, and actions. For example, when you plan to take a walk because you think it is a beautiful day, the inner critic may introduce counter thoughts such as you are not going to really enjoy the walk so why bother. Also, you may have set a goal to embark upon drinking more water, the inner voice may sabotage your efforts by introducing contrary thoughts about the true value of drinking more water so before long you find that you have not been successful in drinking more water and wonder why you did not achieve your goal. In this example, not achieving this goal is directly related to the hindrances caused by the inner voice. The inner critic can also interrupt your thoughts by appearing at a time when you are trying to listen to others, mentally focus, or concentrate on completing a certain task. The inner voice becomes effective at disturbing your concentration by introducing thoughts or ideas that are totally unrelated to the subject you are thinking about or working on and before you realize it the mind begins to wonder and not concentrate on key concepts. The inner voice can also undermine your confidence when you are making decisions or executing a certain action. For example, when you are getting dressed the inner voice may enter thoughts that suggest that the outfit you are wearing is too colorful, too casual, too dressy, or not professional enough for an event that you are attending. Also, when you are authoring a chapter for writing a book, the inner voice may try to convince you that the content is not good enough, the editors may not like the manuscript and that no one is going to buy the book. The damaging thoughts generated by the inner voice can prevent you from achieving your ultimate goal of successfully completing the manuscript for a new book. Because the inner critic is so powerful and can hinder one's productivity, it is vital to develop strategies for silencing this critic and minimizing its effect on one's resulting productivity and performance.

To effectively silence the inner critic one must be intentional about implementing effective strategies to thwart these thoughts before they negatively affect our behavior and resulting actions. In working with various coaching clients, three strategies for effectively controlling the inner critic were discovered. These three strategies are as follows: (1) actively ignoring the inner voice; (2) refocusing the mind to silence the inner voice; and (3) responding to the inner voice with information to refute or rebuke the ideas of the inner critic.

The strategy of actively ignoring the inner critic involves mentally not paying attention to the judgmental thoughts or comments. Implementing this strategy requires the mental discipline of simply not allowing your attention to be diverted to the judgmental thoughts. For example, one coaching client realized that whenever she sat down to complete her goal planning tasks she would have judgmental thoughts by the inner voice that implied she was wasting her time and had more important things to do with her time. It was discovered by this coaching client that this is a very powerful strategy because after a few minutes of actively ignoring the judgmental counterproductive thoughts they diminished and eventually disappeared. Thus she was able to continue with her task and be very productive in getting the planning tasks accomplished.

The strategy of refocusing the mind to stop the inner voice from introducing counterproductive thoughts involves one actively dismissing the thoughts and reengaging the mind in the task or action being performed. To successfully implement this strategy several of the coaching clients have commented that they have had to actively stop and practice deep breathing or some other technique to refocus their mind. Also, some coaching clients have practiced simply pausing and closing their eyes and then reopening them to refocus. Likewise, some clients report that taking a 3- to 5-minute break and then returning back to the task has also helped to silence the inner voice while trying to focus and complete tasks.

Additionally, some coaching clients actively use the strategy of responding to the thoughts of the inner critic by speaking back to it with information that refutes or rebukes the judgmental comments. For example, if the internal critic states that you are not disciplined enough to successfully complete writing a book manuscript, one client states that she would respond with examples of when she has been disciplined such as when she successfully completed assignments or tasks. This strategy of speaking back to the inner voice works best if data-driven examples with facts and figures are used as counter examples.

Prior experience with various coaching clients reveal the importance of persons actively using a strategy to silence the inner critic; thereby, controlling its effect on their performance and productivity. Actively developing strategies for silencing the critic will aid in ensuring that you are prepared when these judgmental thoughts present themselves as you are embarking on important tasks. Please complete the exercise that follows by identifying strategies you will actively use to silence the critic and enhance your performance and productivity.

EXERCISE: Silence the Critic

To actively silence the critic, I will utilize the following strategies:

1.
2.
3.

1.9 CHASE THE FAT RABBITS

As we think of the many task demands faced each day, it is critical to strategically determine which tasks to pursue for ensuring the greatest productivity. There will always be more to do than we have the time and energy to pursue so it is vital to develop an approach for identifying which tasks are most important to pursue. Over the many years of working with various coaching clients, we began referring to the most important tasks to pursue as "The Fat Rabbits." Thus the concept of "Chasing The Fat Rabbits" refers to pursuing those tasks that are most important to achieving your overall goals, objectives, and targets.

Also "The Fat Rabbits" are the tasks with a high impact on achieving targets that have been set for your goals. For example, if you are in sales this may mean pursuing the sales accounts or clients that have the greatest potential to significantly grow your portfolio. If you are in engineering, this may mean working on the tasks that have the potential to significantly enhance your product, service, or system that you are responsible for designing. If you are in management, this may mean completing the tasks that will allow you to broaden the skills or capabilities of the persons you manage to increase their productivity and performance.

Additionally, "The Fat Rabbits" are those tasks that have a higher reward or payoff when accomplished. When thinking of how to use your time wisely, it is always advantageous to think of the potential return to be gained from investing time to work on specific tasks. Knowing the outcome that will be realized is an important consideration to ponder prior to embarking upon a task. By analyzing the situation and determining the results to be gained you can make a calculated decision of whether to embark upon certain tasks or endeavors. For example, if you are in HR team and must give a training class to all employees, perhaps "The Fat Rabbit" task is to develop this course for electronic delivery so that it can be reviewed anywhere and anytime on demand for employees; thus minimizing the time it would traditionally take to complete this task. Likewise if your goal is to significantly increase your protein intake, then perhaps it would behoove you to find one protein supplement that has the maximum nutrients than to have multiple protein supplements or complete one workout that gets the maximum results as oppose to pursuing multiple workout plans. Likewise, if you are in the profession of fundraising it may be wise to invest your time in pursuing a large seven figure donation than pursing many smaller contributions.

In addition, "The Fat Rabbits" are those tasks that have a synergistic relationship to other tasks and allow you to accomplish multiple tasks with less energy. This concept is similar to the old adage that many of us refer to as "killing two birds with one stone." For example, if you have a goal of writing a research journal article and writing a research grant proposal, it would be wise to select one topic that can be used for both the article and proposal. Also, if you are responsible for providing a training course on corporate culture and corporate priorities, then it behooves you to combine the employee orientation course to include both of these topics. Likewise if you work in banking and get paid a commission on consumer lending accounts as well as mortgage accounts, then it is strategic to work with each individual client to simultaneous set up loan accounts as well as mortgage accounts.

As discussed "Chasing The Fat Rabbits" refers to strategically identifying and completing those tasks that significantly advance your productivity or performance. "The Fat Rabbits" tasks are those that upon their completion minimize the total amount of time necessary for realizing attainment of a major goal. Also, "Fat Rabbits" are those significant tasks that can be done to complete a goal using less energy or effort. Likewise, "Fat Rabbits" are those tasks that reduce the amount of steps or actions needed to reach a significant goal. Additionally, "Chasing The Fat Rabbits" means pursuing the completion of the critical tasks that help to cut costs and lessen the amount of money expended to achieve a desired outcome. Finally, "Chasing The Fat Rabbits" refers to prioritizing your plans to focus on completing the tasks which can significantly improve the quality of your efforts and activities; thereby, influencing how well a task is completed. By "Chasing The Fat Rabbits" you can reduce the time, energy, and costs associated with reaching your highest goals and priorities, while increasing the quality of your efforts. Given the rewards to be gained from "Chasing The Fat Rabbits," it is critical that you plan which tasks to pursue with great thought, prior to embarking upon the pursuit of any tasks that must be executed. Therefore your strategic decision to "Chase The Fat Rabbits" will greatly impact your productivity, how much you get completed, as well as your performance, a measure of how well you complete these tasks.

EXERCISE: Identifying The Fat Rabbits

Review the list of your goals, objectives, and targets, list the three most significant tasks (i.e., the fat rabbits) that must be achieved to realize these goals, objectives, and targets:

1.10 CONCENTRATE ON COMPLETING AND CLOSING MATTERS

In sales there is a common phrase of "ABC: Always Be Closing" which emphasizes a key point for those in that field. In this context, this phrase has become a philosophy which pushes each person to remember that the most important activity of the sales cycle is to close which seals and completes the deal. Similarly, the most important activity in ensuring your highest level of productivity and performance is to make sure that you work toward closing and completing various tasks and initiatives. Often we focus on beginning tasks, which is typically the phase that one enjoys most because of the excitement of beginning something new and different. While we may find excitement in the initial phases of projects and activities, we must push ourselves to be mindful of the importance of finishing such tasks or initiatives as well.

There are various factors to overcome when ensuring that tasks and activities are being completed. One factor involves overcoming delays resulting from waiting on others to complete a task which inhibits your efforts to finish tasks. It is often wise to develop a planned schedule of activities with delivery dates and due dates and agree upon following this schedule with everyone involved in the project, program, or initiatives. In addition, we commonly find ourselves waiting on finances, materials, supplies, or other resources necessary to complete the tasks. Implementing strategies of consistently checking with others to ensure that you receive items pertinent to your task completion is essential to your overall success. Also, creating incentives for others to meet key delivery dates are very helpful to ensure overall success. Additionally, being in scenarios where you are waiting on direction from someone else so that you may successfully finish your tasks can be disheartening and cause delays in your productivity. Thus strategies for keeping others accountable and ensuring that they provide you with essentials needed to successfully complete your task must be employed. This typically means establishing meetings, incentives, or reprimands that can be used to get what's needed to advance the completion of the desired tasks or objectives. There are other factors that impede progress and task completion such as procrastination, indecisiveness, and perfectionism. Often, we suffer from busyness which allows other priorities to interrupt the process of completing a task. The power of focus is needed to overcome these hindrances by restructuring your thoughts, plans, and schedule to direct the effort and energy needed to complete the task. The satisfaction of completing the task is actually more exhilarating than the feelings of beginning a task or starting something new.

There are numerous examples of individuals who have started important activities but failed to complete them as planned. Take a moment to think of some tasks that you have started but not yet completed. Think of the time, energy, mental activity, and other resources that have been utilized yet to no avail. The tragedy of not closing and completing matters is that one becomes acclimated to not finishing matters and does not realize that this can become habit-forming. Thus if one is not careful they adopt a pattern of beginning things but not finishing tasks or initiatives. Hence their productivity is undermined. Also their performance begins to degrade as well. The sense of disappointment begins to set into one's mind.

Another factor that contributes to this phenomenon is that of interruption. We experience daily interruptions that interfere with our plans to get things completed. Interruptions such as email, instant messages, casual conversation, meetings, phone calls can significantly impact our productivity; research shows that if we have days where we are constantly interrupted and cannot complete tasks that are important we become frustrated, angry, disappointed in ourselves and these emotional reactions may began to steal our hope in getting things completed. Once we find ourselves in this vicious cycle we begin to think "why bother and why worry" and we begin to alter and lessen our productivity and performance expectations for ourselves. This is dangerous because we lower the expectations for ourselves which ultimately erodes our productivity and performance standards.

Overcoming this cyclic effect involves:

1. Recognizing that it is happening is key;
2. Setting expectations that push you to focus on completing tasks daily;
3. Developing strategies for ensuring that you get task not just started but completed;
4. Developing daily practices that reinforce the habits of completing tasks and closing out initiatives; and
5. Establishing accountability partners and measures to ensure daily task completion.

Exercise: List three strategies you will use to ensure that you are concentrating on completing and closing tasks each day:

CHAPTER 2

Practices for Improving Productivity and Performance

Once strategies for advancing your productivity and performance have been developed, it is critical that you implement practices to remain highly productive and ensure outstanding performance such as those that are discussed in this section as well as others you may be aware of or currently include in your daily routine. Thus this section is designed to provide input on practices that can be developed or further enhanced to support your goals of increasing productivity and performance.

2.1 ESTABLISH A PARADIGM FOR ENSURING CONSISTENCY

While working with my coaching clients, we found that more often than not in life, many of them have not accomplished all of the desires and goals that they have for their life. In fact, when I first began working with many of them, we commonly discovered that frequently they only accomplished 20%–30% of the tasks that they desired to complete weekly. The pattern of working hard for the week, yet only accomplishing a maximum of 30% of what was important, left many of them feeling sad, frustrated, upset, and dismayed. After continuing to work with the coaching clients over a period of time, I noticed that many of them began to achieve as much as 80%–90% of their important task weekly. Upon reflection we discovered that the difference between accomplishing only 20%–30% and as much as 80%–90% of your weekly tasks was a function of implementing practices that ensure consistent behavior and actions. Thus consistency was discovered to be the key. For example, those clients who regularly scheduled a day and time to plan task action items for the week were more likely to actually sit down and develop the action plan required for success. Those who simply resigned themselves to saying, "I will do planning this week" but did not identify a day and time to complete this action item were far less likely to actually get it done. Likewise, those individuals who adopted a paradigm of consistently executing tasks on a certain day at

Key Productivity and Performance Strategies to Advance Your Career.
DOI: https://doi.org/10.1016/B978-0-12-799956-2.00002-4

a standard time were highly likely to get their tasks done within the allotted time. Thus among the hundreds of coaching clients it was discovered that establishing a paradigm of consistency for scheduling a set time to perform tasks was directly proportional to successfully accomplishing the desired plans. Taking the time to establish a detailed plan of how you will consistently use your time to work on specific desires and goals is critical to your success. Fig. 2.1 shows an example of one coaching client's weekly schedule. Notice this weekly schedule was designed to ensure that time is allocated for those desires and goal-related tasks that are most important to the coaching client. While this schedule looks extremely detailed, it did not take long to develop. The average time to develop this type of plan is approximately 45–60 minutes. Making the time to create such a plan will help to ensure that you accomplish 80%–90% of your desired goals and outcomes.

The most difficult step of creating a paradigm for consistency is to develop a detailed schedule as shown in Fig. 2.1 to use as a plan for allocating your time. Once this schedule plan is completed, you will be well on your way to enhancing your productivity and performance. However, it is important to stay on track with following your schedule. It will take a few weeks of practice following the schedule before it begins to become routine or a habit so remember to remain focused on being consistent in following the schedule to ensure that you form the right habits during this timeframe. Also, it will take practice to remain on track. For example, one may have to develop some visual cues to remind yourself or print out a copy of the schedule and refer to it frequently so that you are clear on which tasks you should be working on at a certain point in your day.

Time of Day	Day of the Week						
	Monday	Tuesday	Wednesday	Thursday	Friday	Saturday	Sunday
5–6 a.m.	Devotion	Devotion	Devotion	Devotion	Devotion	sleep	sleep
6–7 a.m.	Workout-Run	Workout-Bike	Workout-Run	Workout-Bike	Workout	sleep	sleep
7–8 a.m.	Workout-Run	Mrng. Prep	Workout-Run	Mrng. Prep	Workout	sleep	sleep
8–9 a.m.	Mrng. Prep	Mrng. Prep	Mrng. Prep	Mrng. Prep	Mrng. Prep	sleep	sleep
9–10	email	email	email	email	email	Tennis/Workout	Mrng. Prep
10–11	Research Proposal	Research Proposal	Research Grants	Research Grants	Research Grants	House/Yard work	Mrng. Prep
11–12	Research Proposal	Research Proposal	Research Pubs	Research Pubs	Research Pubs	House/Yard work	Church
1–2 p.m.	Marketing & Diss	Research Proposal	Research Pubs	Research Pubs	Research Pubs	House/yard work	CHurch
2–3	Marketing & Diss	Prof/Lab development	Authorship	Authorship	CAFÉ to dos	email	FAMILY, FRIENDS, FUN
3–4	Marketing & Diss	Prof/Lab development	Authorship	Authorship	CAFÉ to dos	bills	FAMILY, FRIENDS, FUN
4–5	HWK Dinner Prep	HWK Dinner Prep	HWK Dinner Prep	HWK Dinner Prep	Happy Hour-Friends	FAMILY, FUN	FAMILY, FRIENDS, FUN
5–6	Dinner	Dinner	Dinner	Dinner	Happy Hour-Friends	FAMILY, FUN	Husband Date Night
6–7	Dinner	Dinner	Dinner	Dinner	MY Wild Card Time	FAMILY, FUN	Husband Date Night
7–8	Authorship	Authorship	Kickboxing	Authorship	MY Wild Card Time	FAMILY, FUN	Husband Date Night
8–9	Reading/Relaxation	Reading/Relaxation	Reading/Relaxation	Reading/Relaxation	MY Wild Card Time	FAMILY, FUN	Planning/email
9–10	Bed PREP	Bed Prep	Bed Prep	Bed Prep	MY Wild Card Time	FAMILY, FUN	Reading/Relaxation
10–	Bed	Bed	Bed	Bed	MY Wild Card Time	FAMILY, FUN	BED Pre/BED

Figure 2.1 Example of weekly time allocation plan.

It is ideal if you have the opportunity to follow your schedule uninterrupted. However, occasionally something may cause you to need to alter or modify your plan, such as if you need to travel or have a new task or activity that interrupts the plan. Also, occasionally you may find that you need to change the order of tasks to introduce variety into your week. If you have a slight interruption in the weekly plan, the best option is simply not to worry about changing your weekly plans and get back on the scheduled tasks as soon as you have completed the unplanned activity or task. For example, if you have to attend a company networking event on Wednesday evening from 7 to 9 p.m. then you would simply get back on the weekly schedule plan beginning Thursday morning. Also, if you find that you have to travel to a meeting on Wednesday and Thursday, then you would simply begin following your planned schedule again on Friday if you feel that you do not have any pressing deadlines related to the activity areas that were missed on your schedule for Wednesday and Thursday. However, if you had some pressing activities related to the Wednesday and Thursday weekly plan, then you would modify your schedule for the remainder of the week to insert time for completing the critical tasks that you missed while traveling. For example, since you missed the opportunity to utilize the time scheduled on both Wednesday and Thursday for authorship and there is no other time allocated for authorship in the remaining days of the week, you may want to simply insert the authorship tasks on Friday in the time slot allocated as "My Wild Card Time" or any other time slot where you have flexibility on Saturday or Sunday.

Additionally, to ensure that you establish a paradigm for consistency in accomplishing your planned desires and goals you will have to implement practices to minimize interruptions and distractions. For example, if you are someone who is distracted by instant messages, you may have to turn those off or minimize the dialogue window so that it is not within your view. Also, you may need to consider the practice of turning your telephone on silent or positioning it out of sight while working so that it does not capture your attention. If you are someone who has frequent interruptions or are easily distracted you may also consider inserting a block of time in your weekly schedule that you designate as "time for catching up or completing unfinished task." Establishing a designated time for performing tasks necessary to catch-up on areas where you are behind will help to ensure that you are caught up on getting the urgent and most important tasks completed each week. Also, earmarking time for completing unfinished tasks will help to thwart the problem of ending your week without completing critical tasks. Having this special block on time in your schedule plans will help to ensure that the week ends with you feeling productive and remove undesired stress and frustration.

At the beginning of your week, it is important to insert time into your schedule for weekly planning which will afford you the opportunity to review your schedule and make adjustments for new activities, travel, meetings, and unmet obligations. It is also critical to insert 5–10 minutes each day at the beginning for a review of your schedule to identify any changes needed to ensure daily productivity and high performance.

2.2 FOLLOW A CRITICAL PATH ANALYSIS FOR TASK EXECUTION

Occasionally, some goals will consist of very complex tasks that need to be completed to fully accomplish the goal. The practice for successfully completing complex tasks lies in following a critical path approach for task execution. A critical path is commonly defined as "the sequence of stages determining the minimum time needed for completing an operation." Thus it is important to determine the best order for executing the tasks to reduce rework and redundancy and ensure that the complex task is completed in the shortest amount of time possible. For example, developing a research proposal is a goal that has very complex tasks associated with its completion. Prior to embarking upon completing any task, it is essential to spend some time thinking of the vital steps to perform and the best order for performing these steps to minimize delays or an inefficient use of your time. For example, if you begin writing the research proposal without first reading the preparation guidelines, it is highly likely that you may have to redo the work or add information that was omitted because you were not sure which information was required prior to beginning to draft the document. Likewise, if you are working on a complex task that requires input from others or historical data to successfully complete the task, often it is important to get such information quickly before spending time independently working on the task because this input may significantly alter the scope of work or proposed direction of the work to be done in successfully completing the task. An example

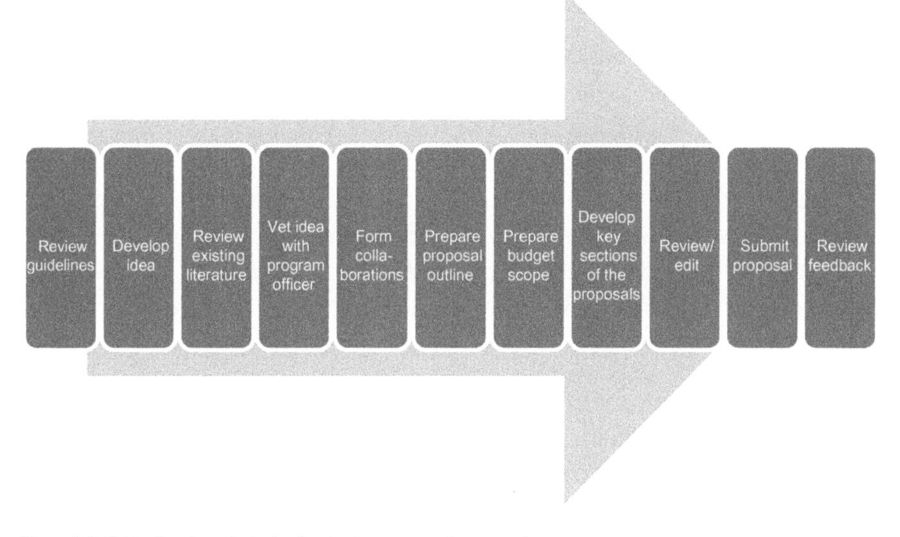

Figure 2.2 Critical path analysis for developing a research proposal.

of a critical path analysis for completing the task of writing a research proposal is shown in Fig. 2.2.

Making time to develop the critical path analysis prior to embarking upon a complex task is time well spent for a variety of reasons. For example, the critical path analysis will present a clear picture of what steps need to be done pertaining to the task. The critical path analysis will also show the best order for executing all tasks which is important to ensure that no steps are omitted and all are executed in the optimal sequencing. This critical path analysis is also helpful in determining how involved or difficult it will be to finish complex tasks. This analysis will expose those areas and subtasks that need additional attention or may be somewhat problematic to complete. This analysis will also identify the points in the process where assistance and input from others are needed to successfully achieve the overall goal and objective. Likewise, the critical path analysis will be advantageous in estimating how long it may take to fully complete all of the tasks pertinent to achieving the overall goal. This information will assist in identifying a timeline for completing tasks and can identify the appropriate starting point to begin work to ensure that you finish by the desired deadline. Adopting the practice of creating a critical path analysis for complex tasks pertinent to achieving an overall desire or goal is instrumental to increasing your overall performance and productivity.

2.3 PRACTICE EFFICIENT ALLOCATION OF RESOURCES

While working to enhance your overall performance and productivity, it is wise to focus on the efficient allocation of resources. In broad terms, resource allocation refers to thinking strategically about how to use available resources to achieve your goals and desires. This concept will help you in developing a plan for allocating your resources among various tasks and actions that you are required to complete. In maximizing your performance and productivity, it is wise to think of the factors shown in Fig. 2.3 before embarking upon task-related action items.

Applying the concept of resource allocation in project management is very important because it provides a clear picture regarding the amount of work that has to be done. By using the concept of resource allocation you can develop a schedule which will help in allocating the right amount of time to completing projects or tasks. Detailed resource allocation practices allow one to plan and prepare for the project's implementation which ultimately translates into achieving desired goals. Using the concept of resource allocation will also assist you in thinking holistically about factors that influence your productivity and performance. Additionally, by implementing the practices of resource allocation, you will develop a thorough understanding of the

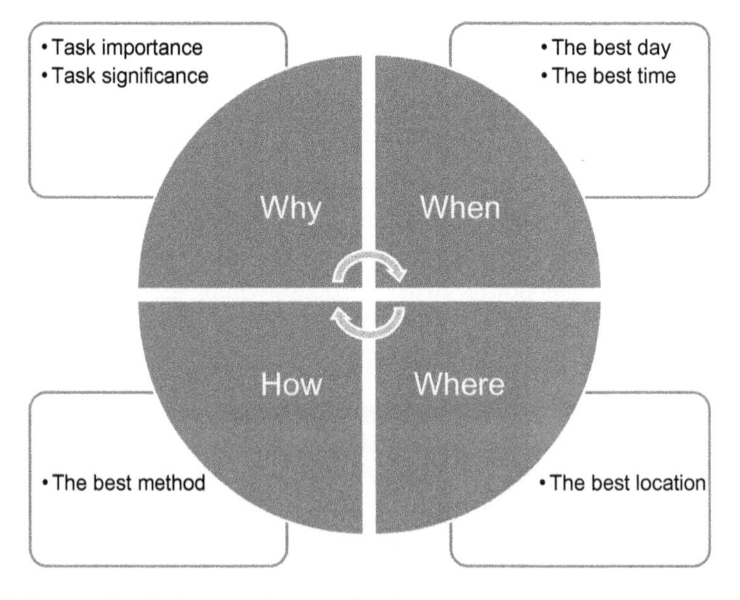

Figure 2.3 Resource allocation factors pertinent to productivity.

importance of establishing routines that enhance your likelihood of being productive.

Likewise, by using resource allocation theory, it is also possible to analyze existing threats and risks to task or project completion. Additionally, applying resource allocation practices in project management helps to manage the workload associated with getting certain tasks completed. By thinking about how to best allocate your resources you will avoid the problem of "over-allocating" yourself. It is important not to over-allocate yourself so that you avoid experiencing burnout and having your productivity drop significantly. When applying resource allocation theory to enhancing productivity and performance, it is important to consider the most pertinent components of human resources. Over the years of working with many coaching clients, we identified several important factors that serve as critical components of human resources that should carefully be considered when planning a task schedule. The most important resources are best explained using the "T.I.M.E.D model" shown in Fig. 2.4.

Fig. 2.4 illustrates the importance of considering various components of the human resources responsible for completing certain goals,

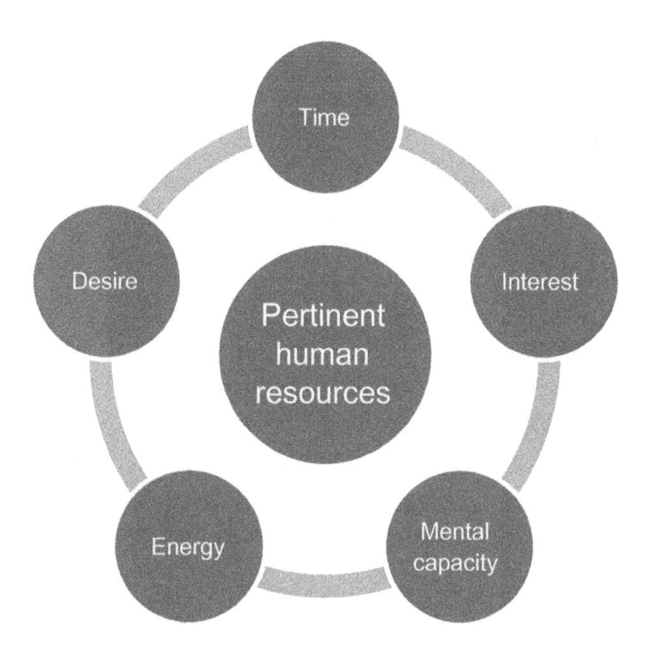

Figure 2.4 Pertinent human resources associated with productivity and performance.

objectives, and targets necessary to achieve overall success. Thus one's productivity and performance are closely tied to the amount of quality time allocated and spent working on a task. For example, if one does not intentionally allocate time to work on a task, then it is highly likely that the task will not get done. Also, if not enough time is allocated to complete a task, then it also follows that one may begin the task but not have sufficient time to complete the task. In working with coaching clients, we observed that it is wise to allocate the majority of one's time each week to working on achieving their most important goals. If less time is allocated, then it is highly likely that significant progress toward meeting these goals will not occur as hoped.

Also, as shown in Fig. 2.4, it is also wise to consider planning task execution based on the interest you have in the task or activity. If your interest in working on the task is low, then you must schedule the task for the time of day or time of the week when you are most disciplined and focused on finishing the task. For example, some people are best at doing their most difficult tasks first thing in the morning or at the beginning of their work day; others have found that their focus on finishing tasks grows stronger toward the later part of their work day because of their desire to finish things before the day ends and so will schedule tasks with the lowest interest toward the end of the day.

In addition to time and interest, as shown in Fig. 2.4, often mental capacity is an important factor to consider when developing your plans for task execution. In working with many coaching clients, we have discovered that one's mental state directly impacts their ability to be productive. For example, perhaps you have noticed times when it is difficult for you to think quickly or pay attention to material being read or listen closely to verbal discussions. Thus points of consideration regarding your mental state's peak periods of performance should be considered when constructing your weekly or daily schedule. For example, if you have the task of writing a research proposal and you know that your mental state's peak performance occurs midmorning, then perhaps it would behoove you to schedule proposal writing from 10:00 a.m. to 12:00 p.m. in your day to take advantage of your time of mental peak performance.

The amount of energy available to work on desired tasks and goals is another important consideration when planning to ensure maximum productivity and performance. In working with coaching clients, it was discovered that one should schedule the task with the expected longest duration of completion during the times when they have the most energy available. Also, if working out and fitness is part of your plan, then it should be scheduled at the point in the day when you have the greatest energy. Also, practices that increase your energy should be used wisely to enhance performance as well. For example, tasks can be divided into components that can be completed within an hour. Also, task allocation plans can be made where you work on a task for 20- to 30-minute durations then switch to another type of tasks. For example, work on the computer for 30 minutes then make an important related phone call, then switch to answering emails for a specified period of time or perhaps read and review pertinent documents. Also when considering the energy resources available, it is important to implement practices that assist in helping the mind and body to rest and recover appropriately while working which helps to ensure that you remain highly productive throughout the day. For example, many coaching clients have successfully implemented the concept of microbreaks throughout the day. A microbreak can be as simple as taking a 2- to 5-minute break during each hour of working. Implementing this practice assists in maintaining the energy level needed to support sustained work activity.

The factor of "desire" within the T.I.M.E.D. model shown in Fig. 2.4 refers to our strong feeling of wanting to work on a task or action item. This feeling of desire can significantly impact our productivity and performance because we naturally have an affinity toward working on tasks that we have strong feelings toward. For example, if we are writing a book and have a strong desire to get it finished, we will make the time in our schedules to work on completing the book. However, if we are supposed to write a book and there is not a strong desire to work on the task, then we are likely to procrastinate and find excuses not to work on the tasks related to the book; thus sabotaging achievement of the goal. Therefore it is advantageous when we have a strong desire to pursue a task; consequently, when possible we should embark upon tasks that are consistent with our desires. However, it is also incumbent upon us to develop practices motivate us to complete tasks when we do not have the desire to do so. In working with my coaching clients, we have discovered that this can be very difficult to do. Thus I advise you to develop some practices that may help you to work on these types of tasks as well. Some coaching clients have used information on the importance of the task and the significance of its associated goal to stimulate their desire in embarking upon completing of the task.

2.4 EMPLOY TECHNOLOGY TOOLS TO EXPEDITE TASK EXECUTION

While working with various coaching clients, we discovered that those who employed the use of various technology tools to assist them with task execution were more consistent and overall successful in achieving their desired goals, objectives, and targets. Specifically, taking advantage of using technology, automated processes, standardized approaches, checklists, templates, and training materials have proven to be very helpful to many of my coaching clients. Implementing practices that utilize various technology tools to expedite task execution can significantly enhance your overall productivity and performance.

Technology advances have significantly changed today's work environment and the manner in which we get tasks accomplished. There are state-of-the-art tools that assist in planning work tasks to be completed such as the various planners and organizers. Also, there are tools that assist in creating schedules that contain information on meetings and other task-oriented information that continues to help keep us updated on task expectations and requirements such as google calendar and various others. Additionally, there are electronic tools such as to do lists, journals, etc., that allow us to keep at our fingertips a listing of tasks to be completed daily, weekly, monthly, and annually. These tools store critical information and provide options for updating such information quickly and seamlessly. Developing practices to take advantage of using these technology to plan, store, and organize information pertinent to executing tasks that are important to achieving your overall success is essential to your productivity and performance. For example, the various features of the google calendar function will allow you to establish a routine schedule for working on various tasks and it will keep track of your success. Likewise, the google calendar features will allow you to set reminders that you can use as prompts to begin working on critical tasks that must be completed. Many of these tools have features that will also assist in collaborating and delegating tasks to others, which helps to simplify the process of assigning tasks to others for completion. It is important to your overall productivity that you explore the use of technology-assisted tools and select a tool that will allow you the flexibility to plan, organize, and track task completion.

Also, it is important to take advantage of technology tools that enable you to be productive while you are on the go and away from a standard office or computer work station. There are applications that can be placed on your mobile devices that allow for flexibility and agility to complete tasks at any time or place. For example, many of my coaching clients have word processing applications on their mobile phones so that they may write books, memorandums, letters, journal publications, and grant proposals at any time. Many take advantage of using these tools to minimize the effects of being idle while waiting during various appointments. Also, many of them use dictation applications that minimize the time to develop a document or create an important message. Also, it has been shown by some of the coaching clients that messaging tools are more efficient for information sharing of time-sensitive task matters than our traditional technique of phone calls or emails because they reduce the time to receive a response. There are various additional examples of technology-aided solutions that will help to bolster your productivity and performance so it is important to carve out a few hours to explore and identify which practices may work best for you.

Exercise: Employing Tools to Expedite Task Execution

1. *List some technology tools that you can employ to help you plan and organize yourself to be more productive:*
2. *List some technology tools that you can employ to assist you in completing more tasks daily, weekly, or monthly:*

2.5 ADHERE TO PRACTICES THAT PROMOTE REST, REJUVENATION, AND STRESS RELIEF

After working with hundreds of coaching clients, we discovered that adhering to rest and rejuvenation practices were vital to ensuring long-term productivity and performance. Also, many individuals found that the key to sustaining their level of productivity was dependent upon implementing practices that helped them to relax and rest appropriately. In addition, many discovered the importance of employing practices that rejuvenated them after periods of high performance; thereby, propelling them to continue working on important tasks related to their desired goals and objectives.

Getting ample rest is critical to maintaining the level of cognitive, emotional, and physical energy needed to be productive each day. Fatigue and tiredness can significantly impact your time, interests, and desire to work on various tasks throughout the day, week, or month. One's level of fatigue or tiredness has been known to negatively affect his or her level of productivity. For example, as one's level of fatigue or tiredness increases then the amount of time that it takes to complete a task increases. Also, as an individual's level of fatigue or tiredness increases their level of concentration decreases causing delays, mistakes, and other performance degradations. Given the vast implications associated with not getting sufficient rest, it is important that each individual work diligently to adopt practices that ensure ample rest. Individual coaching clients have adopted various practices such as deep breathing, meditation, relaxation exercises, visual imaging exercises, listening to white noise, reading, and counting exercises to assist in getting ample rest.

In addition to getting ample rest, many of my clients have discovered that it is critical to implement practices for rejuvenating theirselves for continuing to embark upon completing important tasks related to achieving their overall goals and desires. Specifically, they have implemented practices for replenishing their interests, excitement, and engagement in pursuing tasks that need to be completed. Additionally, many have infused practices that grow their level of joy, excitement, and enthusiasm such as taking a brief walk during the work day or a long run during the day. Some have opted to supply theirselves with tools that provide fun interactions such as stress balls, desktop basketball games, golf putting greens, making paper airplanes, race cars, and sandboxes. Past experience with my coaching clients has shown that those who proactively create a plan for rejuvenating themselves are more likely to remain highly productive.

Likewise in addition to identifying practices to promote rest and rejuvenation, many of my coaching clients have discovered the importance of identifying practical techniques for reducing stress that can be easily implemented in your work environment and life in general. Stress can cause concern, anxiety, confusion, irritation, frustration, and many other emotional responses that can hinder performance and productivity. Therefore it is critical that individuals take a self-inventory to increase their awareness of situations and circumstances that create stressful situations or increase their stress level. For example, after self-exploration, one may discover that time demands, increased workload responsibilities, poor communication with coworkers, scarcity of resources, etc., increases their level of stress. Once, these stressors are identified, practices for reducing the occurrence of these factors or limiting the effect of these stressors should be identified and implemented consistently to avoid negative repercussions on productivity and performance.

EXERCISE: Practices for Rest, Rejuvenation, and Stress Relief

1. *List the practices that you will employ to get ample rest:*
2. *List the practices you plan to implement to rejuvenate and recharge:*
3. *List the factors that serve as stressors for you:*
4. *List the practices you will implement to mitigate stress:*

CHAPTER 3
Conclusion

Achieving the highest level of productivity and performance requires the use of well-known strategies and proven practices such as those discussed in this book. Also, it is wise to remain focused each day on the importance of working toward achieving your goals, objectives, and targets. If one does not remain focused it is easy to become distracted and lose sight of the important tasks that need to be completed daily, weekly, monthly, and annually to advance your productivity and performance pursuits. Actively using the strategies discussed in this book will help you to remain prepared, focused, and poised for success. The key strategies include (1) setting clear goals, objectives, and targets to be pursued; (2) preparing a plan for accomplishing tasks; (3) remaining strategically immersed while performing tasks; (4) employing decisive decision-making to guide your task-centric efforts; (5) adopting a get things done mentality so that you minimize procrastination; (6) controlling the concerns created by thoughts of fear, doubt, and worry; (7) overcoming the limiting views in your mind that sabotage your efforts; (8) silencing the critical inner voice that thwarts your initiative to pursue tasks; (9) chasing the fat rabbits in life; and (10) continually concentrating on completing and closing task matters.

In addition to using the strategies discussed in this book, many individuals have found it invaluable to implement proven practices that boost their productivity and performance. For example, it is key to invoke the following practices: (1) establish a paradigm for ensuring consistency in completing tasks; (2) follow a critical path approach for ensuring the optimal execution of tasks; (3) determine the most efficient allocation of resources to accomplish your desired goals; (4) employ the use of technology and automated tools to expedite task completion; and (5) utilize practices that promote rest, rejuvenation, and stress relief.

While utilizing the strategies and practices discussed in this book to enhance your productivity and performance, it is wise to avoid some

Key Productivity and Performance Strategies to Advance Your Career.
DOI: https://doi.org/10.1016/B978-0-12-799956-2.00003-6

common mistakes made by others. For example, some persons have gotten started but quit too soon to realize the benefits. Remember that it takes a minimum of 21 days to form a habit and it may take as long as 8 weeks to begin to permanently modify one's perceptions, attitudes, and beliefs. Thus give yourself a trial period of at least 3 months to implement the strategies and practices discussed in this book to realize results. Also, focus on augmenting the information presented in this book with your answers to the exercise questions, this will help you to customize the advice to fit your individual needs and performance style. As new challenges or hindrances to your productivity and performance present themselves, expand upon the strategies presented and the practices suggested to develop solutions that eradicate these problems as quickly as possible to avoid sabotaging your efforts to maximize your performance and productivity.

Remember that your overall productivity and performance is a cumulative effect of what you do each day to plan, organize, and execute task assignments. Thus consistency and commitment are very important to long-term success. However, if you find yourself offtrack do not give up and abandon the approaches learned simply find the strength and courage to motivate and inspire yourself to get back on track with your planning and execution. I guarantee that if you consistently follow the strategies and practices included in this book, high productivity and performance results will be your reward and you shall achieve your desired goals, objectives, and targets!

PRODUCTIVITY/PERFORMANCE LOG AND JOURNAL

PRODUCTIVITY AND PERFORMANCE LOG

Date	Tasks Completed	Tasks NOT Completed	OTHER

Future Action Items Planned to Enhance Your Productivity and Performance:

PRODUCTIVITY AND PERFORMANCE JOURNAL EXERCISE

Date:____________________

Today, I rate my productivity and performance level as : Poor, Fair, Good, Excellent

(select one)

The Following Strategies that I used worked well:

____

____

____

____

____

____

The Following Strategies that I used DIDN'T work well:

____

____

____

____

____

____

What changes should I make to maximize Productivity and Performance?

____

____

____

____

____

____

PRODUCTIVITY AND PERFORMANCE LOG

Date	Tasks Completed	Tasks NOT Completed	OTHER

Future Action Items Planned to Enhance Your Productivity and Performance:

PRODUCTIVITY AND PERFORMANCE JOURNAL EXERCISE

Date:____________________

Today, I rate my productivity and performance level as : Poor, Fair, Good, Excellent

(select one)

The Following Strategies that I used worked well:

____

____

____

____

____

____

The Following Strategies that I used DIDN'T work well:

____

____

____

____

____

____

What changes should I make to maximize Productivity and Performance?

____

____

____

____

____

____

PRODUCTIVITY AND PERFORMANCE LOG

Date	Tasks Completed	Tasks NOT Completed	OTHER

Future Action Items Planned to Enhance Your Productivity and Performance:

____

____

____

____

____

____

____

PRODUCTIVITY AND PERFORMANCE JOURNAL EXERCISE

Date:______________________

Today, I rate my productivity and performance level as : Poor, Fair, Good, Excellent

(select one)

The Following Strategies that I used worked well:

____

____

____

____

____

____

The Following Strategies that I used DIDN'T work well:

____

____

____

____

____

____

What changes should I make to maximize Productivity and Performance?

____

____

____

____

____

____

PRODUCTIVITY AND PERFORMANCE LOG

Date	Tasks Completed	Tasks NOT Completed	OTHER

Future Action Items Planned to Enhance Your Productivity and Performance:

____
____
____
____
____
____
____

PRODUCTIVITY AND PERFORMANCE JOURNAL EXERCISE

Date:______________________

Today, I rate my productivity and performance level as : Poor, Fair, Good, Excellent

(select one)

The Following Strategies that I used worked well:

The Following Strategies that I used DIDN'T work well:

What changes should I make to maximize Productivity and Performance?

PRODUCTIVITY AND PERFORMANCE LOG

Date	Tasks Completed	Tasks NOT Completed	OTHER

Future Action Items Planned to Enhance Your Productivity and Performance:

____

____

____

____

____

____

____

PRODUCTIVITY AND PERFORMANCE JOURNAL EXERCISE

Date:____________________

Today, I rate my productivity and performance level as : Poor, Fair, Good, Excellent

(select one)

The Following Strategies that I used worked well:

____
____
____
____
____
____

The Following Strategies that I used DIDN'T work well:

____
____
____
____
____
____

What changes should I make to maximize Productivity and Performance?

____
____
____
____
____
____

PRODUCTIVITY AND PERFORMANCE LOG

Date	Tasks Completed	Tasks NOT Completed	OTHER

Future Action Items Planned to Enhance Your Productivity and Performance:

PRODUCTIVITY AND PERFORMANCE JOURNAL EXERCISE

Date:______________________

Today, I rate my productivity and performance level as : Poor, Fair, Good, Excellent

(select one)

The Following Strategies that I used worked well:

____
____
____
____
____
____

The Following Strategies that I used DIDN'T work well:

____
____
____
____
____
____

What changes should I make to maximize Productivity and Performance?

____
____
____
____
____
____

PRODUCTIVITY AND PERFORMANCE LOG

Date	Tasks Completed	Tasks NOT Completed	OTHER

Future Action Items Planned to Enhance Your Productivity and Performance:

____

____

____

____

____

____

____

PRODUCTIVITY AND PERFORMANCE JOURNAL EXERCISE

Date:____________________

Today, I rate my productivity and performance level as : Poor, Fair, Good, Excellent

(select one)

The Following Strategies that I used worked well:

The Following Strategies that I used DIDN'T work well:

What changes should I make to maximize Productivity and Performance?

PRODUCTIVITY AND PERFORMANCE LOG

Date	Tasks Completed	Tasks NOT Completed	OTHER

Future Action Items Planned to Enhance Your Productivity and Performance:

PRODUCTIVITY AND PERFORMANCE JOURNAL EXERCISE

Date:____________________

Today, I rate my productivity and performance level as : Poor, Fair, Good, Excellent

(select one)

The Following Strategies that I used worked well:

____
____
____
____
____
____

The Following Strategies that I used DIDN'T work well:

____
____
____
____
____
____

What changes should I make to maximize Productivity and Performance?

____
____
____
____
____
____

PRODUCTIVITY AND PERFORMANCE LOG

Date	Tasks Completed	Tasks NOT Completed	OTHER

Future Action Items Planned to Enhance Your Productivity and Performance:

PRODUCTIVITY AND PERFORMANCE JOURNAL EXERCISE

Date:______________________

Today, I rate my productivity and performance level as : Poor, Fair, Good, Excellent

(select one)

The Following Strategies that I used worked well:

____

____

____

____

____

____

The Following Strategies that I used DIDN'T work well:

____

____

____

____

____

____

What changes should I make to maximize Productivity and Performance?

____

____

____

____

____

____

PRODUCTIVITY AND PERFORMANCE LOG

Date	Tasks Completed	Tasks NOT Completed	OTHER

Future Action Items Planned to Enhance Your Productivity and Performance:

PRODUCTIVITY AND PERFORMANCE JOURNAL EXERCISE

Date:______________________

Today, I rate my productivity and performance level as : Poor, Fair, Good, Excellent

(select one)

The Following Strategies that I used worked well:

____

____

____

____

____

____

The Following Strategies that I used DIDN'T work well:

____

____

____

____

____

____

What changes should I make to maximize Productivity and Performance?

____

____

____

____

____

____

REFERENCES AND RESOURCES

The following references and resources are available to assist you in continuing to advance your efforts towards enhancing your productivity and performance.

[1] V. Burton, Successful Women Speak Differently, Harvest House Publishers, 2016. 978-0-7369-5681-9. <www.harvesthousepublishers.com>.

[2] The Center for Advancing Faculty Excellence at Tennessee State University. <www.theacademiccafe.com>.

[3] L. Freeman, The New Boss: The Guide to a Fabulous Lifestyle, Tate Publishing and Enterprises, LLC, 2016. 978-1-63418-164-8. <www.tatepublishing.com>.

[4] M. Gadsden-Williams, CLIMB: Take Every Step With Conviction, Courage, and Calculated Risk to Achieve a Thriving Career and Successful Life, Open Lens/Akashic Books, 2018. 978-1-61775-624-5. <www.akashicbooks.com>.

[5] C. Johnson, D. Susan, Too Blessed to Be Stressed: Words of Wisdom for Women on the Move, Thomas Nelson, 1998. 0-7852-7070-1.

[6] Project Management strategies listed at <www.Projectmanager.com>.

[7] S. Toler, Minute Motivators for Leaders: Quick Inspiration for the Time of Your Life, 2014, Harvest House Publishers, 2015. 978-0-7369-6821-8. <www.harvesthousepublishers.com>.

[8] The Science of Happiness: New Discoveries for a more Joyful Life (2017). Time Magazine Special Edition. <www.timemagazine.com>.

INDEX

A

Always Be Closing (ABC), 35

B

Beliefs, 1, 27, 30, 53–54

C

"Chasing The Fat Rabbits", 33–35
Coaching clients, 9–12, 15–16, 18–20, 28–29, 31–33, 39–40, 46–47, 49–50
Cognitive approaches, 1
Commission errors, 19–20
Commitments, 13–14, 24, 54
Consistency, 24, 54
  paradigm for ensuring, 39–42
Critic, 29–33
Critical path analysis for task execution, 42–43

D

Decision
  decision-making, 21
  table, 22–23, 23*t*
  theory, 22
Decisive decision-making techniques, 21–23

E

Enthusiasm, 28, 50
Errors, 19–20

F

False Evidence Appearing Real (FEAR), 26
"Fat Rabbits, The", 33–34

G

Goals, 46–47, 50. *See also* Long-term goal
  plan for accomplishing, 9–17
  setting, 1–9
  short-term, 2–4
Google calendar function, 48–49

H

Human resource components, 45–46

I

Incidence rate of errors, 19–20
Indecisiveness, 35–36
Inner critic, 29–32
Internal critic, 29–32

L

Long-term goal, 2–4, 4*f*
  addition of targets to listing, 11*f*
  example listing of objectives for, 8*f*
  template for developing list of objectives, 9*f*

M

Mental capacity, 46
Mind, limiting views in, 26–29

O

Objectives, 1–9
Omission errors, 19–20

P

Perfectionism, 35–36
Performance journal exercise, 55–56, 58, 60, 62, 64, 66, 68, 70, 72, 74
Performance log, 55, 57, 59, 61, 63, 65, 67, 69, 71, 73
Planning process, 4–6, 9–17
Procrastination, 35–36
Productivity and performance, 2–4, 15–16, 54
  Chasing The Fat Rabbits, 33–35
  completing and closing matters, 35–37
  critic, 29–33
  decisive decision-making, 21–23
  fear, doubt, and worry, 24–26
  goals, 1–9, 16
  key strategies, 53
  limiting views in mind, 26–29
  mentality, 23–24
  objectives, 1–9
  plan for accomplishing goals, 9–17
  practices, 53–54
    adhering to practices for rest, rejuvenation, and stress relief, 49–51
    allocation of resources, 44–48

Productivity and performance (*Continued*)
critical path analysis for task execution, 42–43
establish paradigm for ensuring consistency, 39–42
strategic immersion, 17–21
technology tools to expedite task execution, 48–49
targets, 1–9

R

Rejuvenation, practices for, 49–51
Resource allocation, 44–48
Rest, practices for, 49–51

S

Scheduling system, 9–12
Self-esteem, 9, 27
Self-image, 9, 27
Self-value, 9, 27
Self-worth, 9, 27
Short-term goal, 2–4, 3*f*, 5*f*, 14*f*
addition of targets to listing of priority 1 areas for, 10*f*
example listing of objectives, 7*f*
template for developing list of objectives, 8*f*
Strategic immersion, 17–21
Stress relief, practices for, 49–51

T

Target setting, 1–9
Task
execution
critical path analysis for, 42–43
employing technology tool to expedite, 48–49
objectives, 15–17
task-oriented information, 48–49
Thoughts, 1, 17–18, 30–31
of fear, doubt, and worry, 24–26
judgmental, 31–32
T.I.M.E.D model, 45, 47–48

CPI Antony Rowe
Chippenham, UK
2019-02-17 21:00